EINSTEIN
DEFIANT

Selected other titles by Edmund Blair Bolles

The Ice Finders: How a Poet, a Professor, and a Politician Discovered the Ice Age
A Second Way of Knowing: The Riddle of Human Perception
Remembering and Forgetting: Inquiries into the Nature of Memory
So Much to Say: How to Help Your Child Learn
Galileo's Commandment: An Anthology of Great Science Writing (editor)

EINSTEIN
DEFIANT

Genius versus Genius in the Quantum Revolution

Edmund Blair Bolles

Joseph Henry Press
Washington, DC

Joseph Henry Press • 500 Fifth Street, NW • Washington, DC 20001

The Joseph Henry Press, an imprint of the National Academies Press, was created with the goal of making books on science, technology, and health more widely available to professionals and the public. Joseph Henry was one of the founders of the National Academy of Sciences and a leader in early American science.

Any opinions, findings, conclusions, or recommendations expressed in this volume are those of the author and do not necessarily reflect the views of the National Academy of Sciences or its affiliated institutions.

Library of Congress Cataloging-in-Publication Data

Bolles, Edmund Blair, 1942-
  Einstein defiant : genius versus genius in the quantum revolution / by Edmund Blair Bolles.
    p. cm.
Includes bibliographical references.
  ISBN 0-309-08998-0 (hbk.)
  1. Quantum theory—History—20th century. 2. Physics—Europe—History—20th century. 3. Einstein, Albert, 1879-1955. 4. Bohr, Niels Henrik David, 1885-1962. I. Title.
  QC173.98.B65 2004
  530.12'09—dc22

                                    2003023735

Printed in the United States of America.

To Kelso Walker
and the rest of the crew, volunteers all.
Tanzania XII: U.S. Peace Corps, 1966-1968

Bohr asked me to sit down and soon started to pace furiously around the oblong table in the center of the room. He then asked me if I could note down a few sentences as they emerged during his pacing. Bohr never had a full sentence ready. He would often dwell on one word, coax it, implore it, to find the continuation. This could go on for several minutes. At that moment the word was 'Einstein'. There was Bohr, almost running around the table and repeating: 'Einstein . . . Einstein . . .'. After a little while he walked to the window, gazed out, repeating every now and then: 'Einstein . . . Einstein . . .'.

—Abraham Pais, *Niels Bohr's Times*

Contents

A RADICAL FACT RESISTED

1	The Opposite of an Intriguer	3
2	Not German at All	12
3	I Never Fully Understood It	26
4	Independence and Inner Freedom	35
5	A Mercy of Fate	51
6	Picturesque Phrases	57
7	Scientific Dada	68
8	Such a Devil of a Fellow	76
9	Intuition and Inspiration	86
10	Bold, Not to Say Reckless	94
11	A Completely New Lesson	102
12	Slaves to Time and Space	110
13	Where All Weaker Imaginations Wither	116
14	A Triumph of Einstein Over Bohr	127

A RADICAL THEORY CREATED

15	Something Deeply Hidden	137
16	Completely Solved	146
17	Exciting and Exacting Times	162

18 Intellectual Drunkenness   184
19 The Observant Executrix   195
20 It Might Look Crazy   204
21 Taking Nothing Solemnly   211
22 How Much More Gratifying   218

A RADICAL UNDERSTANDING DEFIED

23 Sorcerer's Multiplication   229
24 Adding Two Nonsenses   236
25 Admiration and Suspicion   241
26 An Unrelenting Fanatic   248
27 The Secret of the Old One   251
28 Indeterminacy   256
29 A Very Pleasant Talk   264
30 The Dream of His Life   276
31 The Saddest Chapter   285
32 A Reality Independent of Man   293
33 A Certain Unreasonableness   299

Afterword   305

Bibliography   309

Sources   317

Index   337

Part I

A Radical Fact Resisted

1

The Opposite of an Intriguer

In 1918 Albert Einstein's face did not yet capture the living union of tragedy with genius. His hair had yet to shoot crazily from his broad, serious brow, and his eyes did not yet laugh and x-ray at the same time. He was still decades away from sticking out his tongue at a camera. Instead, photos from that period show a man not quite 40 years old who seemed firmly settled into middle age. His hair had gone gray; his eyes had lost some sparkle that had been evident in pictures taken only four years earlier. Anybody who saw him the day he rode a tram to the Reichstag probably took him to be just one more exhausted burgher looking for food in a starving city.

Outside Einstein's tram window, Berlin was cracking open like Humpty Dumpty. Until recently people had thought the Great War was going well. The Russians had surrendered and the Kaiser's troops along the western front had again begun to advance toward Paris. "Victory all along the line" had been the watchword, and then suddenly World War I was over. Germany cried uncle and the whole imperial system collapsed like a handsome marionette whose puppeteer has been felled by a stroke. One flinch and down it went.

There had still been some dreamers. The industrialist Walther Rathenau had wept when the war began, but now he appealed to the people to "rise in defense of their nation." The generals knew such pleas were hopeless and ignored him. Germany's navy mutinied; soldiers on the Russian front had been radicalized by Bolshevik propaganda, while masses of troops on the western front began deserting. The police had disappeared from sight. Rumor alone was in charge.

Max Wertheimer also rode aboard Einstein's tram. Remarkably, he looked younger than he had back in 1914. The Kaiser's army had forced him to shave off his great black beard and take on a soldier's air. Wertheimer, a Czech-born Jew, was Prussian in neither outlook nor manner, but he had served Germany ably enough to win an Iron Cross. Einstein had been among the war's strongest opponents and shocked even other pacifists with his treasonable opinions that favored Germany's outright defeat. Yet the beardless soldier and the long-haired peacemonger were good friends, united by a common delight in ideas. Both men knew what it was like to be seized by a notion and not to rest until the puzzle was resolved. Once, in 1910, Wertheimer had been riding a train when he suddenly had a new idea. Descending at the next station, he checked into a hotel and began experimenting. A week later he emerged with the blueprint for what came to be known as Gestalt psychology. Then he boarded another train and continued his journey.

That had happened a world ago, before rumor had gained command, before a whisper had been force enough to remove the Kaiser from his throne. In a desperate effort to stop a revolution, the emperor's chancellor had said in a telegram that "the Kaiser and King has resolved to renounce the throne." It was a lie. The Kaiser still dithered about whether to resign or stand firm, but such indecision did not matter once Berlin's streets heard about the telegram. The Kaiser was finished. He fled his capital, lucky to have avoided the last tsar's fate. The German throne took its great fall. The architect Walter Gropius was on furlough from the Italian front and saw a Berlin crowd insulting officers. *This is more than just a lost war,* he told himself, *a world has come to an end.*

The city that Einstein saw rushing by his streetcar window was paying the monstrous price of having lived on a tangle of lies. Berlin still looked whole. It had seen no enemy aircraft (as London had) and felt no artillery shells (as Paris had), yet Berlin had become a desperate, starving, refugee-filled city of dismay. All that had previously seemed real had proved false as a dream. And now, adding to the torment of sudden damnation, a plague was taking liar and prophet alike. Sweeping around the world, the flu had arrived in town, killing 300 Berliners a day and turning every little November cough into a reason for terror.

Einstein was bound for Berlin's revolutionary center. When news spread about the emperor's supposed abdication, a crowd assembled outside the Reichstag and demanded to see Germany's leading socialist politicians, who, just then, were surviving the Allied blockade by dining on watery soup. Eventually a politician strolled out to the Reichstag balcony and pronounced a few words, "The old and the rotten," he shouted to the crowd that strained to catch his speech, "the monarchy has collapsed. Long live the new!" Then, getting carried away, he added, "Long live the German Republic!" There was a cheer and the politician returned to his soup. Until that moment, the plan had been to create a new, constitutional monarchy with the emperor's grandson as the new king, but somebody had said *republic* and that was enough smoke and mirror for there to be no more German kings.

A third man rode the tram to the Reichstag with Einstein and Wertheimer. This was Einstein's colleague in physics, Max Born, and he had a cough. Born had been in bed when Einstein telephoned him to report the latest crisis, that revolutionary students had seized the university and were holding some professors and its rector hostage. Einstein thought he might have some influence with the students and asked for Born's support. So faithful Max Born had left his bed and now Einstein, Born, and Wertheimer were riding toward the parliament to see what they could do to rescue the hostages.

That hour, throughout the city—the country even—little groups were scurrying about, trying to save something of the old while revolution created something new. When Gropius realized that a world had come to an end, he told himself that Germany had to find a radical solution to its problems. Many, probably most, Germans agreed, but they hesitated, too. Young boys expected to keep their tattered soccer balls while their parents worried over the few dog-eared marks they had tucked away. Even Rosa Luxemburg, leader of the socialists'

most radical wing and longtime revolutionary, argued to prevent
Leninist terror from becoming Germany's new order.

Einstein was as much a bohemian as Picasso, following the demands of neither civilization nor property, but living out his life from
his own necessity. Since boyhood he had seemed to despise everything
about the old Germany and its authorities. Hatred for the old and the
rotten was a familiar tune for him. He had been born a German but so
loathed Germany's hereditary prerogatives and enforced soldiering that
he renounced his citizenship when he was 16 years old and became a
Swiss. Yet, now that the plumed hats and strutting uniforms had committed suicide, Einstein found that there was something in the old that
he wanted to save after all. He returned to Germany before the war
because, whatever else he might have thought about it, Berlin was
science's capital. A wit during prewar times had said that only 12 people
in the world understood Einstein's theory of relativity and 8 of them
lived in Berlin. The greatest physicist in the world had found that the
fabric of scientific society bent his path right back toward the Prussian
heart. The life of the mind had prospered under the emperor and
Einstein wanted the scholarly life to continue in the new Germany,
whatever the new turned out to be.

But how to save it? Although Einstein often charged at windmills,
he rarely had a plan of action. He saw them spinning and made his
move. His idea that day seems to have been to appeal to the students as
reasonable people who had inadvertently taken innocent men pris-
oner. Einstein's party was not going to ask for the students to surrender control of the university but to press for them to let the professors
and the rector return to their homes.

Seven years earlier a Czech novelist named Max Brod had known
Einstein in Prague and noticed the absurdity of ever sending this particular man to confront a hive of schemers. In a historical novel, Brod
used Einstein as the model for his portrait of Prague's greatest scientist,
the astronomer Johannes Kepler. In it he wrote that Einstein/Kepler
"was the opposite of an intriguer; he never pursued a definite aim and
transacted all affairs lying outside the bounds of his science as in a
dream."

It seems unlikely that any other passengers aboard the tram noticed Einstein. He was not yet world famous and his theory of relativity was generally unknown. Physical discoveries that had made popular splashes were X-rays (the mysterious, invisible something that saw bones through skin), radium (the stone that glowed in the dark), and the theory of the "Solar System" atom with electrons flying like planets around a sun-like nucleus. Einstein had written a supposedly popular account of relativity, but his idea was too specialized and difficult to have attracted much public notice. One promising student who struggled to master it was Werner Heisenberg, at the time in a secondary school in Munich, and even he thought Einstein's account made for tough sledding. Relativity was not anything you could ever expect to seize the public's imagination the way X-rays and radium had.

Berlin's parliament appeared in the distance. The Reichstag was a large, heavy building designed with a huge dome and the massive arches of Renaissance Rome. It had been built in the 1880s as part of the Kaiser's effort to transform backward Berlin into a capital that could rival London. Across the building's front ran the phrase, "To the German people." The Kaiser had hated that slogan—objecting to it as "revolutionary"—and he had resisted having any such sentiment carved into his building, but in 1916, while the Battle of Verdun ground up Germany's young men by the hundreds of thousands, something had to be done to express the solidarity between the people and their government. The revolutionary sentiment had been carved into stone.

Max Born was the most handsome of Einstein's trio. He was trim and looked fit. He was also a veteran, had been Wertheimer's commander in fact, and yet he too was a friend of Einstein's; perhaps he even loved Einstein. Partly it was because Einstein was a champion physicist while Born was merely a great one, and partly it was because Einstein was a very good man, but mostly it was something else. Einstein, in his absentminded way, was a charmer. He had a grace about him that good people appreciated, so they forgave him for not being like them, for not being, as so many chums are, a mirror in which we see and forgive ourselves.

Einstein had been no help to the Kaiser during the war, even though Berlin's most honored physicists enticed him to the city specifically to reinforce Germany's scientific leadership, and even though many close friends had turned their science to waging war. One of Einstein's friends had directed the development and manufacture of mustard gas and other poisons for use in the trenches. Max Born had supervised artillery range-finding technology, and Wertheimer had used his psychological expertise to create sound systems for locating Allied artillery. Wertheimer then expanded on his location techniques to create the first sonar system placed in U-boats. During the Great War, most scientists had made their brains available to their local warring power. (Britain's leading physicist, Ernest Rutherford, had duplicated Wertheimer's antisubmarine work, but, of course, for the Allies' cause.) The days when crowned heads dismissed and choked science seemed long past. The deliberate embrace of stupidity that had once silenced Galileo was replaced by an effort to make science lay its golden eggs for those in power. It took courage to deny a state its eggs.

As Einstein's delegation descended from their streetcar, students wearing red armbands swarmed about the parliament building, enjoying a political lark, pretending that the world made easy sense and that they knew what was what. Einstein and the two Maxes made their way toward the armbands. As men of imagination they inevitably discomforted all claimants to power, but as scientists they were used to being respected. The emperor had fled, but they did not suppose the fury that had deposed a king might be turned their way. At that minute, rioting students were holding their colleagues hostage, yet they did not worry that anybody would suggest they too should be taken under the revolution's wing. Could they recall the scenes from history, a great chemist fleeing through narrow back streets, pursued by a mob that knew exactly whom it was chasing? Years later Einstein shook his head when he reminded Born about the confrontation at the Reichstag. "How naive we were, even as men forty years old!! I can only laugh when I think about it. Neither of us realized how much more powerful is instinct compared to intelligence."

Instinct at first held its ground. Enter the Reichstag? Not possible. Intelligence had no ready retort, but then instinct's arbitrariness showed itself. A newspaperman recognized that it was the great Einstein trying to gain admittance, and he told the student guards to let the group pass. The students cheered the hero-pacifist as he came among them.

Most of the students would not have applauded Einstein in 1914. When the war began, almost all Germany cheered it. The socialist politicians, who had seemed to hate Prussia's aristocrats and generals, voted their proud support for the coming slaughter. Intellectuals cheered as well. The novelist Thomas Mann supported the war with his full throat, and many an ordinary German marched happily away. Back then Einstein's refusal even to cheer the boys as they stepped off in their new uniforms had been shocking, but the battlefield meat grinder, the government's lies, and the sudden defeat on the western front had shown Einstein to be a prophet. The socialists, restored now to pacifism, assumed that surely he was one of their own. Once the student guards realized who he was, they happily waved Einstein and party through their lines.

Inside the building lay signs of the sudden change in national management. Cigarette butts were scattered everywhere and the carpeting was littered with paper, dust, and dirt. A rifle pile lay in the lobby. Attendants dressed in parliamentary livery served as memorials to what the place had been. Einstein's party and their reporter escort entered a meeting to find the most radical students busy declaring themselves enemies of liberty. The party watched in silent dismay as the students passed a resolution that only socialist doctrines should be taught at the university, only socialists should be accepted as professors, and only socialists admitted as students.

After their vote the students looked to the prophet for praise and support. What did Einstein think? Einstein knew too much about efforts to silence professors with unpopular opinions. He himself had been targeted by student protesters when they objected to his pacifism and wanted him thrown off campus, or better yet, chained up as the Brits did with Bertrand Russell. Einstein told the Reichstag students that freedom of thought was German university life's most precious staple, and he was sorry they had voted as they had. Their faces fell. Max Born never forgot their astonished gawks. Their pathfinder was defying them all.

Einstein changed the subject. He had not come to the Reichstag to debate intellectual freedom, but to see about their colleagues' physical freedom. Couldn't they be released? Einstein waited. Would the students yield or would they go against their hero? They eluded the test by stepping back and denying any authority over the matter. They told the trio that they had already passed control of the university to the new revolutionary government. A politician might have said something—*you are revolutionaries; surely you have the authority to act in revolutionary solidarity with one who has long dreamt of this day*—but Einstein turned and led his entourage away. The reporter knew he had a story and still tagged along.

When the trio and its journalist escort made their way into the governmental headquarters at the Reich Chancellery, they entered a world of impossible problems. The discomfited academics whom they hoped to rescue were nulls in comparison with other matters that harassed the government. Peace had not ended the Allied blockade; indeed, armistice had made the blockade more effective, for it destroyed the wartime system for evading the stranglehold. But chaos proved an advantage to Einstein's group. First, they were able to walk in and present their cause to Friedrich Ebert, the head of the revolutionary government. Then, too, Einstein's party had brought a problem that Ebert could actually solve. The blockade, British hatred, French demands for vengeance, the incoming refugees, the flu, and the hunger had no solutions, but Ebert could sign a bit of paper ordering the release of the academic hostages.

Leaving the Reich Chancellery with their *ad hoc* document, Born recalled, the threesome exulted "in high spirits feeling that we had taken part in a historical event, and hoping to have seen the last of Prussian arrogance ... now that the German democracy had won." They moved on to the university, clutching Ebert's signature like a medieval messenger bearing a white flag.

If Einstein were handed a paper like the one he presented to the students, he might have wondered whether there was anything behind it, but the university's occupants lacked Einstein's profound ability to doubt the unsupported, and the paper did its work. In the end, people need to point to something and say, "There, that's my authority." Einstein pointed to a chit and, in confused Berlin, nobody could trump him with a counterchit, so a scrawled paper signed (allegedly) by Friedrich Ebert proved authority enough.

With their colleagues released, Einstein, Born, and possibly even the ultraskeptical Wertheimer went away delighted with both the world and themselves. Berlin was starving, plague-ridden, and tempted toward bolshevism, but Einstein, having charged a windmill and seen it surrender, could hope that reason and justice were about to triumph.

2

Not German at All

In the midst of Berlin's chaos, a problem entirely of his own making tormented Einstein. As so often happened with Einstein, what he saw as a problem would have struck many others as a solution. Switzerland had offered him an escape hatch from Berlin. Einstein, however, wanted to say No. But the Swiss were so nice and the Germans so barbarous that he hated to be rude and snub a generous offer. He needed a way to say No and Yes at the same time. Einstein had once built a theory on the idea that an object can seem to be both at rest and in motion, so if anybody could find a No-Yes answer he was probably the one. But no solution jumped at him.

There were Berliners who hoped Einstein would stay, but they understood all too well why he or any other person would grab the opportunity to flee. Germany was up for grabs. Strange prophets with odd ambitions had begun to appear even before the war ended, and the moment the Kaiser fled they began pressing their way toward the center ring. On the very first Sunday following the armistice, a man of distinctly proletarian face, with short-cropped hair, creases in his cheeks, and exhaustion beneath his eyes, entered the city's Protestant cathedral. He was Johannes Baader, and he waited calmly in the church's pseudo-Italian setting. The ornate decor would surely have horrified Martin Luther—Saint Martin, the cathedral's artwork seemed to make him. Then, as the pastor was about to begin his sermon, Baader rose and shouted out, "Just a minute. I'm asking you what Jesus Christ means to you. . .".

He had more to say but already there was a commotion. Baader was seized and silenced. The authorities wanted to charge him with blasphemy, but the speaking notes he carried demanded only greater respect for Jesus. The next day Berlin's newspapers were full of tales of Baader's "insult" to propriety. Baader was one of those strange prophets, a member of the Dada movement, trying to rescue Berlin.

Dada was provocateur art that sought to replace "bourgeois" Europe with the ideals of international art. Its aims cannot be defined more precisely. The previous summer, in a famous statement of enthusiastic contradiction, the Berlin Dadaists had issued a manifesto that concluded, "To be against this manifesto is to be a Dadaist!"

That was Dada's version of a Yes-No reply, and if Einstein thought he could get away with saying, "To reject Zürich's offer is to accept it," he probably would have tried. Might the land of strict accounting and precision clocks accept such an answer? Well, Zürich was Dada's birthplace. Guarded by the Alps, neutral Zürich had long avoided the fates of Paris and Berlin, cities on the plains where armed warriors found easy maneuver. Lenin and James Joyce were Switzerland's most famous anti-war refugees, but many anonymous ones had come there as well to express, under authority of their own souls, their visions. Some anti-warriors with artistic imaginations had, at a *boîte* called the Cabaret Voltaire, begun a protest movement that spread to the warring capitals on both sides of the battle lines. When the war ended, Dada and communism were the only radical movements found in both the conquered and conquering nations. Communism was already becoming a Soviet tool, but Dada remained a decentralized, international movement. It marked postwar Germany's encounter with romantic irrationality.

One of Dada's earliest statements boasted, "The Cabaret's role is to remind us that, beyond the war and nationalities, there are independent men who live by other ideals." It sounds like a movement Einstein could have joined, and, indeed, he was as independent and as anti-national as any Dada poet. Einstein saw science the way Dada's proponents saw art, as an alternative to military madness, and like the Dadaists he hated nationalism. "What is truly valuable in our bustle of life is not the nation," he said, "but the creative and impressionable individuality, the personality." Dadaists across Europe took that doctrine of personal authority as truth's one sure point. Einstein, too, believed in a community of hard-toiling laborers at imagination.

Yet Einstein and Dada could never have made common cause. The Dada Manifesto, issued in Zürich during the summer of 1918, proclaimed, "Logic is complication. Logic is always wrong. It draws the threads of notions, words, in their formal extension, towards illusory ends and centers. Its chains kill; it is an enormous centipede stifling independence." Einstein understood their point. He had been engaged in deeply creative thinking for perhaps 20 years and had produced a series of wonders—showing how to count atoms with a schoolhouse microscope; abolishing the notion of an elusive, universal substance called the ether; abolishing, too, the ancient belief in absolute space and absolute time; redefining how light interacts with matter; redefining gravity—and he liked to say that his creative work had not been logical. "Invention is not the product of logical thought," he insisted, and a Dadaist would have understood, but then Einstein added, "even though the final product is tied to a logical structure."

Logic for Einstein was not hackwork, not a mechanical procedure for generating information, and definitely not the Dadaists' hated military commander who ordered people where to go. Logic, in Einstein's eyes, was a map showing the limits of the understood. A great explorer like Sir Richard Burton looked at a chart and saw some empty spot, possibly decorated with a fanciful annotation like "Here there be tygers," and he said to himself that's where he would go next. So Burton scouted out Mecca and Lake Tanganyika. For Einstein, the tygers were places where logic had no more to say. His greatest papers often began by noticing a paradox, a point where logic contradicted itself. Then, just as a composer resolves two contradictory themes by inventing a unifying symphony, Einstein would unify his physical themes behind a new language for describing the world. *Yes/no.* What was the music that would unify them?

When the Great War ended, radiation was the uncharted territory facing physics. Newton's mechanics had considered matter—the heavy stuff that falls when you drop it, sits still when you set it down, and rolls on forever if you hurl it through empty space. In his early years Einstein had done radical work on matter, but he soon saw that the most revolutionary mysteries had to do with light, heat, energy—all those immaterial enigmas that physicists lumped together as radiation. As science's boldest explorer, Einstein was hoping to unite radiation and matter under one grand theory, but when it came to science, he was not a lunatic adventurer, not one of those mad Englishmen who went gloriously off the map only to discover that the team should have brought warmer clothing. Einstein knew that he needed to understand more about radiation and what it was.

In his younger days when he wanted to understand some profound mystery, Einstein would retreat into the temple of his mind until finally he could reemerge with a prophetic message. The rest of the physics world would gape, wonder if the revelation could be true, and then discover that Einstein had announced something so real that he had birthed a whole new branch of physics. How did Einstein do that? The goings on in the temple were hidden behind a veil and Einstein's method was as undiscoverable as Shakespeare's or Michaelangelo's. But by the war's end Einstein was not quite so invisible. The next dozen years would reveal the great scientist struggling openly with a great mystery. As the story of an individual those years would show how Einstein made war on his own ignorance. On a larger scale it showed how scientific imaginations in general wrestle with nature. On a still larger and even more general level it showed how difficult and wondrous it is for every sort of dreamer to create new things, and why reason is part of any fruitful imagination.

For Einstein the war had exposed unreason's price and he had no sympathy for those groups, whether they called themselves artists or patriots, who were trying to organize the postwar under one of irrationality's many banners.

"Say yes to a life that strives upward by negation," the Berlin Dada Manifesto called, "Affirmation: negation: the gigantic hocus-pocus of existence fires the nerves of the true Dadaist—and there he is, reciting, hunting cycling—half Pantagruel, half Saint Francis, laughing and laughing!" In short, exalted Dada, Yes-No.

How different was the speech made that revolutionary November by Berlin's leading scientist, Max Planck. Bald and wiry, he appeared as sober as a bank examiner, yet there must have been something unconventional about him, for years earlier (in 1900) he had discovered radiation's strangest mystery, the quantum. Now, in late 1918, while revolution tugged at Germany, Planck traveled past the Brandenburg Gate to the Prussian State Library—another of Berlin's grand buildings completed just in time to greet the war's outbreak. The Prussian Academy of Sciences met there and Planck had come to encourage the assembled savants to face the future with courage and confidence. Everyone in the room knew that behind the words, Planck's own heart was broken. One of his sons, Karl, the one whom Planck had considered worthless and without purpose, had been killed at the front. It tore Planck's soul to know he had never recognized his son's value until it was lost. A few months later Planck had lost a daughter, Grete, from complications following childbirth. Even so, he kept a sober face before the Prussian Academy, telling it, "If the enemy has taken from our fatherland all defense and power, . . . there is one thing which no foreign or domestic enemy has taken from us: the position that German science occupies in the world."

Einstein loved to laugh and his spirit united more of Pantagruel and Saint Francis than his colleagues dared recognize. There was also less of Einstein in Planck's patriotic science than the Academy cared to acknowledge, but when it came time to bet, *rouge ou noir*, on Fortune's spinning wheel, Einstein could be counted on to slip all his chips beside Planck's. Einstein would never have phrased it so belligerently, but he agreed with Planck on German science's importance. That was part of why he wanted to say Yes to Berlin, if only he could avoid saying No to Zürich.

Already, however, a new theme sounded softly in the back of the orchestra; Germany might have begun to say No to Einstein, No to science. A new book published the previous summer perfectly suited the new pessimism and the public's doubts about reason. This was Oswald Spengler's *The Decline of the West*, and the book was beginning a bestsellerdom that would take it through 30 printings in five years, and then another 30 printings following a new edition in 1923. The book inspired constant comment among intellectual groups. Its gloomy doctrine held that European civilization—what Spengler called *Faustian Culture*—had passed its maturity and was entering an inevitable decline into senility and death.

Spengler's word "Faustian" said baldly that the West had sold its soul to the beast in return for power. Western claims for science as eternal truth were vanity. Science, preached Spengler, was just as much the cultural moment's product as painting or music, and at the Faustian lie's heart sat the illusion that logical, fixed laws defined reality.

Einstein agreed that the war had exposed many lies. He was eager that still more of them should be publicized so that everyone could understand how false were the idols that had taken millions of their sons and brothers. Yet he was no nihilist; he believed with the confidence of a pope that the world makes sense and that it can make sense to us mortals. His ambition was the mystic's quiet prayer: to find the face of What Is.

Before the war Berlin had been the best place for that pursuit, but now Europeans of all classes and education loathed Germany and Germans. Across the continent good people were shaking their heads over what to do about friends in Germany. In Holland the patriarchal figure of physics, Hendrik Lorentz, a slender old man with a vital face, drafted a letter to a Belgian patron asking,

> What should our attitude be towards the Germans? The misery and suffering they have caused worldwide, the injustices and atrocities committed by their government and their armies, rightly abhorred by all decent people, has as you know, made a very deep and painful impression on me. In addition, I can understand perfectly well that for the time being, Belgian and French scientists want nothing more to do with them. . . . And yet, if we are talking about Germans, we really mustn't lose sight of the fact that individually, they are all different. A man like Einstein, a great and profound physicist, is not 'German' at all, in the sense of the word as it is used at the moment; his own judgment on the events of the last few years wouldn't be any different from yours or mine. . . . All things considered, I feel I must suggest to you that we shall not formally exclude the Germans; in short, we should not close the door on them for ever.

That *for ever* gives the game away. Lorentz knew it was going to take some time before Germans were allowed back into the arms of international science.

Einstein knew it too, even sympathized with the attitude. Prussians were never his choice neighbors. He once told a friend, "These cool blond people make me feel very uneasy; they have no psychological comprehension of others. Everything must be explained to them very explicitly." And then late in the summer, just before the war's end, he was offered an escape via an unusual joint professorship at both Zürich University and Zürich's polytechnic school. Einstein had graduated from the polytechnic and had taught previously at the university. He considered Zürich his true hometown. After he had renounced his German nationality (in 1896) he saved his pocket money until (in 1901) he could afford to apply for Swiss citizenship. Democratic, free Switzerland seemed to him the model for what a country should be. Besides, in Zürich he could work without the distractions of the allied blockade, the revolution, and the desperate starving people.

Einstein wanted no distractions. Physics had become as confused as German politics and it demanded a clear head. Newton's and Maxwell's great achievements had been reconsidered and, while not forced into abdication, placed under a new constitution. A mystery had been swallowing physics for almost 20 years now. At first Einstein had been perhaps the only one alive to realize what was happening, the way a sailor in the crow's nest might be alone in noticing a storm thickening beyond the horizon. Max Planck had discovered the quantum, but Albert Einstein was the first to say aloud that the new quantum theory of radiation was revolutionary.

The mystery that was eating centuries of thought was as old as the dreams around campfires. When prehumans huddled beside a flame, listening to zebras barking out in the darkness, just what were they gathering round? Why did they feel warmer near the fire? Why could they see one another's faces? If they knew, they forgot to pass along the answers to their children. The Greeks buried the problem by naming fire as one of the world's basic building blocks. Its heat and light were just the way things were. Poets and mystics often use fire to express a duality, spirit, and matter, a soul that lives in the world and yet touches something invisible. When chemistry replaced alchemy, Lavoisier proved that fire is simply the rapid union of oxygen with a substance. The fire was the action of the chemical union, but this understanding still left the mystery untouched. What was that burning light? What the heat? What the work that fire did? For thousands of years imaginations had played with the campfire's dream and still the boldest dreamers asked themselves, *What is this flickering thing, really?*

The quantum notion was conceived during an attempt to understand how light is related to temperature. Planck's interests had been abstract and theoretical, but they carried practical implications. Anybody who has ever seen pictures of a steel plant knows that its hot furnaces glow with a brilliant white light. As the flame reaches farther from the source of the activity, its color shifts to yellow shades and then to red. These color changes mark changes in light intensities at different temperatures. Planck derived a formula that linked energy (heat) and light. Using Planck's math a person could, among other things, tell how hot a point in a steel furnace is just by measuring the light waves it gives off. But buried in Planck's formula was the mystery that had shaken Einstein right through.

"What's to be done?" Einstein had written his Swiss friend, Michele Besso, the previous August as he fretted over the offer to return in triumph in Zürich. At that time he was still tempted to just say Yes. "Difficult days of pondering lie behind me." He added, "Proof: I dreamed I had cut my throat with a razor."

Einstein could not expect his neighbors to thank him for risking his throat by staying put in Berlin. Planck would be pleased by his remaining, but few Germans had the mathematical imagination to appreciate Einstein's importance. Most of them, like most of us, were unlikely to be moved by anything, no matter how bizarre, found in an equation, and they would not admire the way Einstein questioned equations as eagerly as poets question clichés.

Einstein's success rested on his talent for finding the concrete meaning behind an equation's abstractions. He had first startled physicists in the spring of 1905 when he published a paper offering visible proof that molecules and atoms really exist. In those days many chemists and a few physicists still considered atoms to be theoretical fictions. Einstein's proof of their reality built on statistical ideas developed by a thick-bearded Austrian named Ludwig Boltzmann. Boltzmann was also a firm believer in atoms and Einstein sometimes wondered why he had not found the proof first. It was a modest man's question. Einstein's creation required a special imaginative energy. Taking one of Boltzmann's statistical equations (one that Boltzmann so loved, he had it carved on his tombstone), Einstein showed how it applied to everyday concepts like time and motion. Until then, Boltzmann's formula had been used only to calculate remote abstractions like probability and entropy in gases. Einstein proved that Boltzmann's abstract idea could be applied to visible things like pollen grains floating in water.

If you look through a microscope to inspect pollen in a water drop, you will see the grains moving in endless jigs and jogs known as Brownian motion. They jump this way and that, getting nowhere, yet never standing still. A botanist named Robert Brown had brought this seemingly perpetual motion to scientific attention in 1828. Did the motion result from something inside the little specks? Or was there something in the water causing the motion? Einstein took the second view and proposed that water molecules knock the pollen about. He made his theory so precise that a person holding a stopwatch and measuring the jigs could calculate the number of molecules in the water. In one short paper Einstein explained Brownian motion, proved the reality of molecules and atoms, and created a lasting technique for analyzing seemingly random fluctuations. Not bad for a patent examiner who couldn't get a job teaching at a university.

Einstein's great strength lay in that ability to find physical meaning in abstract ideas. A scientific "step," as Einstein called these imaginative feats, meant going beyond the received wisdom to show that a fact said something deeper about reality itself. Einstein's strides were the kind of insight long mytholigized in the story of Archimedes leaping from his bath. Supposedly Archimedes shouted "Eureka" when he saw past the banal fact that climbing into a tub causes the water to rise and recognized what was really going on.

It was this search for steps that explain physical surprises that kept bringing Einstein back to Planck's quantum. Up in the crow's nest Einstein had recognized an extra meaning behind Planck's equation. To Einstein, Planck's work was as amazing as it would be for a more ordinary thinker to learn that something fundamental about human psychology could be measured exactly by multiplying a person's IQ by some idealized, fixed batting average. The union is as contrary as Dada's linking Saint Francis with Pantagruel. IQ is brain; batting average is brawn. Granted, brain and brawn might have some connection, but this isn't even your brawn, it is some unchanging, one-size-fits-all batting average. Such an equation makes no sense. So if I told you that I could measure, say, your creativity by combining your IQ with a batting average, you would not take me seriously, but if for the next 20 years I showed you that more and more psychological traits could be measured precisely by multiplying IQs and batting averages you might begin to wonder what on earth was afoot.

Einstein sure wondered. He could have wondered in more comfort in Zürich, but he could not bring himself to leave Berlin. The Swiss offer forced Einstein to face a secret about himself. In Berlin he was admired but stayed always apart, always an outsider. In Switzerland, intimacy was easy, but Einstein finally admitted to his old friend Besso, "Here [in Berlin] everybody approaches me only to within a certain distance, so that life unrolls almost without friction." He had developed a taste for the way the cool blonds left him alone. So it would be Yes to Berlin.

Left alone, Einstein could work on his mystery. Back in 1900 Planck had shown that by multiplying two things that had no business keeping company he could predict astonishingly precise relations between color and temperature. The first multiplier in this case was radiation's frequency, symbolized by the Greek letter v (pronounced *nu*). For visible light, frequency determines color. Technically, the frequency refers to how many times per second something vibrates. This part of Planck's formula held no surprises. It was like the IQ part of our imaginary psychology formula. You would expect intelligence to have something to do with creativity. In the same way, physicists expected measures of light and other radiation to include its frequency. After all, many experiments had proven that light is a wave.

The simplest proof of light's wave nature is to shine colored light, say red light, on two holes to illuminate a screen beyond. If you arrange the holes and screen properly, the red light will include dark lines where no color shines. Light creates these dark lines because the waves coming from the two holes interfere with one another, just as a room with poor acoustics can have dead spots where sound waves interfere with themselves.

The surprise came in the second part of Planck's formula because it measured a discrete chunk of action—a "quantum of action," Planck called it—symbolized by the harmless-seeming symbol h. Planck's number h never changes and is known as Planck's constant. Planck multiplied his h by the frequency of radiation to get what he described as "not . . . a continuous infinitely divisible quantity, but . . . a discrete quantity composed of an integral number of finite equal parts." In other words: when you multiply v and h you do not get something continuous, like a wave, but something discrete, like a particle.

The meaning of hv—a particle-like something emerging from a wave—astonished Einstein. Material things, like stones, come in chunks. Immaterial things, when they are real enough to have any form at all, come in continuous flows, like time. To see the difference between chunks and continuous things, imagine a slingshot that fires a stone into a swimming pool. The slingshot sends the stone flying. When the stone hits the water, however, ripples spread out in all directions. The stone is a chunk or particle and acts in a chunk-like way. It moves in one direction and arrives in one place. The ripples, meanwhile, behave most unchunkishly and flow in all directions, following one another across the pool like cars on a train. These differences result from the way chunks move as a whole while waves spread when they move. So how does a chunkish hv emerge from the pure wave v?

Planck first introduced h as a desperate calculation when no other, more reasonable, energy description worked. He had reached his wits' end searching for a formula and then found that if he used h the formula worked. So, okay, it worked; use it. But then Einstein used hv again, this time in what he called light quanta. Again, the formula worked, but it was so aggressively paradoxical and contrary to everything known about light waves that almost no other physicist in the world took it seriously. Since then, however, the paradoxical hv had turned up again and again. Einstein also used it in a theory about heat, and then a Danish physicist named Niels Bohr used it to analyze the light given off by excited molecules. By the close of 1918 all sorts of basic physics could be measured by using hv. Especially amazing was the way these calculations gave very precise answers. If you multiplied a wave's frequency by Planck's constant, you got an impressively exact, discrete number that, time after time, matched the experimental data.

Einstein called his hv a light quantum, making it a definite thing, an indivisible unit of energy. Others, however, even people like Bohr, who used hv in their own calculations refused to speak of "light quanta." They denied these were actual energy granules bounding through space. A difference like this can seem subtle to the point of invisibility, but Einstein, who believed that scientific concepts identified real things, and Bohr, who wanted only to describe the outcome of experiments, eventually found themselves in a long, slow-stewing battle that forced everyone interested in the nature of science to pick a side.

Even during a political revolution, Einstein could count on Berliners to leave him alone to ponder what hv meant. In Zürich, however, he had a wife and children. Einstein found it easy to say No to family. From early on he had played at love with many women, but he seldom became emotionally entangled with any of them. When his emotions did join in, it was usually a sign that physics had become frustrating, and as physics was rarely frustrating for him, emotional involvement was also rare. Einstein had married his wife, Mileva, during the most difficult time of his life, when nobody in Europe was willing to give him a physics job or even take his letters seriously. Years later, after he had been hailed as a giant, he became tormented by the difficulties in revising his theory of relativity, and again his fancy turned toward a woman, his cousin Elsa. That emotion, too, cooled once he gained sure footing in his wrestle with relativity. By then, however, he had settled in Berlin and sent his wife and children marching back to Zürich.

With relativity resolved, Einstein's attention returned to the quantum. What was radiation and how did it work? Light quanta seemed particularly vexing. It was increasingly clear to Einstein that radiation, at least in its most familiar form, was the emission of light quanta, but light quanta had this puzzling dual nature of wave and particle that made it difficult to say how light behaved. You could perform one set of experiments to show it acted at a specific point, like a particle. A

second set of experiments showed that it moved like a wave. And then
there was Einstein's own genius, always dashing up to the crow's nest
and spotting something new. In 1917, he had written an article on
quantum motion in which quanta appeared to move spontaneously,
that is, without any of the physical causes that Newton's laws require.
Neither the time nor the direction of such spontaneous actions seemed
predictable, except in a statistical way. The whole quantum mystery
appeared to lie on wretchedly incomplete ideas.

Most other physicists had more practical worries. Einstein's question—*What is radiation?*—was the sort of broad puzzle raised by amateurs and three-year-old children. In 1918 a typical physicist was an
experimenter, someone like the young American, Arthur Compton,
who was studying X-ray behavior without paying any attention to
quantum ideas. Even other theoretical physicists asked narrower questions than did Einstein. Einstein's good friend Paul Ehrenfest had been
among the first to join the quantum investigation, but he was ready to
compromise. He wanted "a general point of view," as he put it, "which
may trace the boundary between the 'classical region' and the 'region
of the quanta'." Ehrenfest was willing to accept both classical theory
and quantum radiation, despite the contradictions, if only he could
know when to use which rule. When it came to physics, contradiction
was impossible for Einstein to tolerate.

In his personal life, however, he was more flexible and finally, as
1918 was closing, Einstein dreamt up a way to say Yes to both Berlin
and Zürich. He could not possibly say No to Berlin now that it was
beaten and ruined. Once Germans were despised by the whole world,
they inevitably became more appealing in Einstein's eyes. He wrote to
his friend Lorentz that it was "*a priori* incredible that the inhabitants of
a whole great country should be branded as morally inferior." To the
Swiss as well he said Yes. Unfortunately, he informed them, he could
not abandon his duties in Berlin, but because he greatly appreciated
the offer and "to show his gratitude to his fellow citizens," he would, at
no charge, be delighted to give a semiannual series of a dozen lectures,
starting in early 1919.

Meanwhile, Einstein continued on his own to seek out the law that underlay the world. Already there were stirrings that law had failed and the new world should be based on something else—pick your instinct: power, heroics, terror, art. Einstein was one of the last of the old knights who was still questing to slay the dragons of anarchy and find the law that lies behind life's quietest riddle.

3

I Never Fully Understood It

Once it was settled that Berlin would remain his home and that the Swiss would welcome his free lectures, Einstein grew eager to visit the city where he had gone to school. Accompanied by his cousin-mistress Elsa, he boarded the train at Berlin's Anhalt Station, a giant train shed that had been built on "artistic" lines to evoke power and civilization. Its salmon-pink walls, Roman archways, and circular windows sought to evoke the grandeur that was Caesar's, although perhaps it also expressed the unspoken insecurities behind Prussian ambition.

Once through the station's double porticos, however, the quick-time sounds and oily smells of mechanized travel overwhelmed any nostalgia for the philosophy of Marcus Aurelius. The porters had plenty of luggage to tend. Einstein traveled with a violin, books, paper, and more clothes than gossip might have predicted. His suit material looked cheap, but, perhaps thanks to Elsa, he was a presentable dresser in those days. Photographs show him with various colleagues and always it is the colleague who is the cliché dowdy savant while Einstein looks better. In those days, too, he had a fine, broad-brimmed hat that suggested there might even be a dandy hidden somewhere in that rich soul. Beside him aboard the train, Elsa—short, past 40, older than Einstein—was a lively, good-humored woman who knew many things, including how to leave her cousin alone when he slipped into one of his trances. In their compartment Einstein was free to think deeply and wildly. Many of his examples of relativity were set on trains: *Imagine a train dashing along the tracks at the speed of light when a bolt of lightning strikes the locomotive. . . .*

Even fidgety souls can sink into themselves aboard a train, so you can appreciate how deep Einstein could dive. The critical point to remember, of course, is that Einstein on a train, transfixed by his own daydreaming, is Einstein doing his most important work. "The essential in the being of a man of my type," Einstein said about himself, slipping so naturally into generalization that he replaced even "I" with an abstraction, "lies precisely in *what* he thinks and *how* he thinks, not in what he does or suffers." Reporters and magicians might suspect the old man was trying to force a card on his audience. What he thought was often so much nobler than what he did, but there is no denying that Einstein succeeded in his science because he had a nearly perfect scientific imagination. The most fascinating question about him was always how that imagination worked. The frustration with Einstein is that while we can easily picture him seated on a train—hugging himself slightly for warmth with Elsa seated beside him enjoying a magazine—it is impossible to reverse the view and get inside that large head to see and feel what it was making.

Earlier that year, however, Einstein gave us a peek into what he hoped to achieve from his trances. At a gathering of the Physical Society held in honor of Planck's sixtieth birthday Einstein was supposedly celebrating the way Planck thought, but he generalized from Planck to himself and to all creatures of imagination. He said, "Man tries to make for himself in the fashion that suits him best a simplified and intelligible picture of the world." Whoa! Einstein knew he was overgeneralizing there; knew that many people ignore the world altogether and think exclusively about themselves, or they think not at all and simply act. He knew, too, that dreamers are so much more varied than doers. There are mathematical and scientific imaginations, literary imaginations and visual ones, moral and legal imaginations, criminal and philosophical ones. But while he knew all that, he did not care. He, Planck, Lorentz—the people he most esteemed—were all lost in thought about a simplified, intelligible world. Traveling to Zürich, Einstein was rapt in the presence and search for that simpler, more intelligible picture.

When it came to practical tasks, however, men of his type were as absurd as tigers in tuxedos. They would have done better had they stayed in bed. Indeed, when Einstein did not have lawyers with pressing engagements, he enjoyed his bed's full benefits. He was one of those lucky ones who could sleep until he awoke naturally. Without the daily pressure to be up and out somewhere, he had the creative person's freedom of lying under the covers and playing with the ideas that greeted him when he awoke. The champion lie-abed of those years, of course, was Marcel Proust who passed most of the Great War propped on his pillows, working in a soundproofed room that kept the life of Paris from intruding on his thoughts.

The images Einstein liked to conjure were not scenes from the house where he grew up or the faces of friends, but elegant geometric shapes. In his head he saw an imaginary light that cast shadows while a disk moved across a crystal sphere's invisible surface. Einstein could watch the shadow change its size and shape, and he could recognize what these changes implied about the rules of physics. Playing with images, Einstein had pictured his way to atoms, to relativity, and to light quanta. But even the freest thinker suffers from reality's demands. As 1919 began, both archetypal dreamers, Proust and Einstein, each of whom had passed the Great War imagining a different world, found themselves snagged by the flypaper of circumstance. In Paris, Proust received the galleys to his work's second volume and discovered that, besides thousands of typos, the whole thing had been printed in an unreadably tiny font. Meanwhile, Einstein was due in the Zürich courts, completing his divorce from Mileva.

A few hours out of Berlin, Zürich-bound trains passed through Göttingen, a university town with so illustrious a history in mathematics that even Napoleon spared it from the torch. Einstein had become familiar with Göttingen's quiet glories because, when he was struggling to create his general theory, its mathematicians provided invaluable help. Yet even there Einstein's abstractions were too much for some. Before the war Einstein had given a lecture at Göttingen and, as he explained space-time, many grew restless and one professor had stormed out, muttering "This is absolute nonsense."

That was the challenge the late risers posed for the up-and-at-'ems of the world. New ideas can be grasped only when other people follow along in the creator's imaginative effort. People who really want to experience Einstein's imagination must quit trying to crawl into Einstein's skull and go into their own. Einstein had based his general theory of relativity on a novel geometry that was unknown even to most physicists, and the few people who did know the math still had to work to discover what Einstein saw. It took Lorentz several months to work through the math before he understood the achievement well enough to congratulate Einstein on a brilliant success. Meanwhile Ehrenfest confessed wretchedly to Lorentz that he was still baffled.

Beyond Göttingen the train route to Zürich went west-south-west, diving into territory devastated by actions that had been loosed without imagination. No battles had been fought here. No invaders had arrived, but many of the countryside's young men had been re-placed by a rich crop of widows. Those men who did survive were liable to be armed and angry. Germany was teetering over possible civil war. Its people had been exposed to lessons many bright souls had thought the masses of humanity could never learn. When, before the war, Thomas Mann began writing *The Magic Mountain*, he thought he had a funny idea, a twist on his *Death in Venice* story about a great and mature artist confronting love and death. Suppose, Mann asked himself, the protagonist was not a sensitive aesthete, but an ordinary fellow, a mediocrity, a not very gifted engineering student who had to come to terms with the basics of human existence. In 1913 this thought led Mann to private laughter, but well before the war's end, the conflict had drained away all notion of how such a theme could be funny.

At the war's start every person of action in Europe signed up for the great experience. More, for it was not just Europe that responded. Even in British Kenya and German East Africa the Masai herders were excited to learn that their two overlords were going to war. Only the people most fervently committed to dreaming found something else to do, but by 1918 action alone had been exposed as emptiness. The generals were tired, the people exhausted, the colonized millions, including East Africa's Masai, were disgusted. The generals had begun with plans aplenty but when none succeeded, they could only propose trying the same schemes again and again, hoping each time that events would end differently.

Now that the battles had been lost and won, the world was splitting in two. On one side were those who knew it was time to think again and shape a replacement world. Dreamers, Einstein had said at Planck's sixtieth birthday celebration, try to "substitute" their "simplified and intelligible . . . cosmos . . . for the world of experience and thus overcome it. This is what the painter, the poet, the speculative philosopher, and the natural scientist do, each in his own fashion. Each makes this cosmos and its construction the pivot of his emotional life, in order to find in this way the peace and security which he cannot find in the narrow whirlpool of personal experience." Opposite these creators were the destroyers. They were the ones who preferred to nurse the hatreds bred by four violent years. Already, anonymous veterans, like Corporal Hitler, and glamorous heroes of combat, like Hermann Göring, were furious. They blamed the pacifists and the thinkers for all the sacrificing that had gone for naught. They would have loved to tear down any undefeated survivor of the past.

A few hours beyond Göttingen, the train tracks carried Einstein and Elsa to Switzerland. At the border they showed their passports and were out of Germany. The Swiss were fat, free, and undogged by history. "Brilliant landscape and satisfied citizens, who have nothing to fear. This is how it looks," Einstein reported to his friend Max Born. Zürich was a tidy city along a lakeshore, with the Alps to the far, hazy east. It was as busy and orderly as Berlin was desperate and confused. Then Einstein added this comment for Born, "But God knows, I prefer people with anxieties, whose tomorrow is threatened by uncertainty. How will it all end? One cannot tear one's thoughts away from Berlin; so changed and still changing."

In mid-January, revolution, or really counterrevolution, broke out in Berlin. The revolutionaries, called Sparticists, never had a chance. They were crushed after a few days' fighting. Meanwhile, during those dangerous days, Switzerland's only intrigue came from foreigners, who were caught by the alert police. Germans and Russians were seized as they made their way to Paris to murder the French prime minister, Clemenceau. Then Trotsky's two brothers escaped from a French jail and were seized in Switzerland. Still more Russians and Germans were captured near Lake Geneva, on their way to murder the American president, Woodrow Wilson, at the Paris peace talks.

Some of Einstein's closest friends were in Zürich, waiting for his return. Michele Besso had known Einstein since before his marriage, and, when nobody in the world wished to hire Einstein, Besso had found him a spot in the patent office. Besso was a member of Einstein's "Olympians," young men who got together to laugh and talk about science. Besso became famous as the only person whom Einstein acknowledged in his relativity paper. He loved and admired Einstein, yet he had been appalled by the way his friend treated Mileva, and he served as her intermediary and defender.

Einstein was the sort who is better served by bordellos than by marriage. He occasionally patronized brothels and, years later, grumbled, "Marriage is the unsuccessful attempt to make something lasting out of an incident." His emotions were tied to his dreams, not to his experiences. He could find no lasting satisfaction in being loved, and he did not aim for it. Nor was he in any way concerned about social appearances, so there were no forces pushing him toward domestic stability. When he had first arrived in Berlin its physics community had been delighted at the prospect of working with so eminent a thinker, one who was still in his prime. To their dismay, the city's chemists and physicists were promptly drawn into mediating a matrimonial separation. In Berlin Einstein had immediately resumed a very public affair with his cousin. He began to leave Mileva and the children alone for days at a time. Within months of their arrival in the German capital, Mileva returned to Zürich. Their two boys went with her.

When Einstein reached Zürich he passed January amid a strangle of lawyers who were needed to legalize the death of his marriage. Even then he had science to cheer him. He gave a month-long series of two-hour lectures. Innovative thinkers do not always make great teachers. Planck recalled how he had taken classes from Hermann Helmholtz, Germany's greatest physicist before Einstein. To Planck's dismay, Helmholtz was uninterested in his subject, unable to speak much above a mumble, and unlikely even to get through a calculation at the blackboard without making serious errors. Einstein was different. He liked talking to nonspecialists about physics, and nonspecialists liked listening. He kept his points clear and Einstein kneaded humor in with his numbers and theory.

Einstein's secret was that for him physics was not just a collection of facts, but something transformative that got behind a familiar experience and gave people a way to see the reality behind the show. Many important scientists disagreed with Einstein. They held that the facts of appearance are all there are to truth, and facts-only science did exist. The biological classification system credited to Linnaeus had provided facts without any deeper truth. With the Linnean system naturalists could organize their discoveries according to an efficient, easy-to-remember method based on anatomical similarities. The system served as an excellent device for organizing an ever-expanding body of information, but the categories and relations within them were meaningless and expressed nothing real. Under the Linnean system lobsters and grasshoppers shared a common blueprint while caterpillars and spiders did not. Why? The question was empty and should not be asked. That meaninglessness is why ordinary people find that scientific names inspire the very essence of tedium: *Carcharhinus plumbeus, Galeocerdo cuvier, Carcharias taurus, Squalus acanthias, Lamna nasus.* Such technical blather can make people want to scream, and not in the I'm-having-fun sort of scream that comes while watching a scary shark movie, but an I-can't-take-it-any-more sort of scream. Einstein omitted all that pointless information from his lectures. Facts were nothing to him until they were meaningful.

As passionately as Monet believed in color, Einstein believed in the power of theories to tell us something about the reality behind the surface. He loved taking his audiences that extra step where technical systems become meaningful realities. Einstein preached that the universe is lawful; it works the same way everywhere so that any intelligent creature, no matter where or when in the universe, can find the same laws. For him, abstractions like matter and energy were not just words; they were as real as turnips. Acting on that conviction he had shown that ultimately even matter and energy are forms of one real thing. Did he leap from his bath when he saw the single equation that linked matter and energy? That little equation said there is one thing sometimes seen as energy (E), sometimes mass (m). The one form— mass, say—is equivalent to the other if you factor in the speed of light squared (c^2). Einstein loved this discovery because it dug so deeply. It united so much. It explained so many odd facts.

It is a fact, for example, that the sun shines forever, rising and setting more faithfully than the truest Penelope. It seems like a great fire in the sky except that it never burns out. It is more like a divine fire, like the burning bush that Moses saw. The sun is encased in flames and yet it never consumes itself. Einstein's simple-looking $E=mc^2$, however, got behind the sun's appearance to see its real face. The sun burns eternally because it does not burn the way worldly things do. Somehow it is slowly transforming its enormous bulk (its m of the equation) into an even more enormous (mc^2) ocean of energy. And what do you know? It turns out that the sun will not shine forever. Some day the m will be used up. Stars everywhere are slowly shrinking like melting glaciers, and everywhere they are melting according to the same law. For lay audiences, Einstein's confidence and message was as electrifying as when Charles Darwin took his great stride and showed that the Linnaean classification was not merely an abstract, meaningless tool; it revealed the historical relations between life forms. Lobsters really are close to grasshoppers, and that closeness tells us something of how nature works. By combining surface fact with deeper meaning, scientists had created a powerful tool for exposing the law behind the random play of experience.

Between lectures Einstein finished up with the courts. By coincidence another famous expatriate in Zürich, James Joyce, had a court case at the same time. It would be amusing to pretend a scene where the modern novel's creator and the creator of modern physics pass one another on the court steps without knowing anything of either the other's genius, or pettiness. Perhaps it happened. For sure, on February 14, St. Valentine's Day, 1919, Einstein was officially declared an adulterer and Mileva was granted the divorce that Albert demanded. She retained custody of the children. In the settlement, Einstein agreed that, if he ever won the Nobel Prize, he would give the money to Mileva. Einstein, as the guilty party, was forbidden to remarry for two years; however, Swiss law could not reach into Germany and Einstein

expected to defy it soon by marrying Elsa. Yet his passion for his cousin
had already cooled and he was not moving on to marry at last his life's
great love. Perhaps if he had not been taken ill late in the war and seen
how important it was to have somebody ready to minister to him, he
would never have bothered with a divorce.

Einstein's lectures continued after the divorce. He faithfully
climbed up Zürich University's hillside to tell anyone interested
enough to listen about how the cosmos held together. Some students
asked him to speak about quantum theory, but he told them No.
"However hard I tried," he explained, "I never fully understood it."
The answer must have struck some as ridiculous, like Galileo refusing
to lecture on motion because he did not understand the very revolution he had wrought. Yet Einstein was serious. Neither he nor anybody else had yet cried "Eureka" after taking the imaginative stride
that would transform the quantum hv from a surface fact into a meaningful statement about reality's hidden face. How does an atom seem
spontaneously to emit a chunk of wave? Why does the light quantum
go in one direction rather than another? Nobody yet knew. Einstein,
whether he was in Berlin or Zürich, was never far from his quest to
find out what quanta really were; then he would indeed have a great
lecture to give. His women were not happy; his children were distant,
but the emotional pivot of his life turned on finding how quanta fit
into nature. The puzzle occupied his thoughts when he sat at his desk
and again when he lingered in bed. He was sure that, in time, quanta's
meaning would become apparent. He, or somebody, would get there.
For now, however, Einstein limited himself to lecturing on the realities
that he understood.

Meanwhile, in chilly Copenhagen, Niels Bohr was struggling with
the same puzzle. He had shown that hv could be used to describe
atomic structure, although how the structure worked also remained
mysterious. In Munich, one of physics' best teachers, Arnold
Sommerfeld, was studying the same problem. Sommerfeld boasted that
he had been the first to teach a class in relativity and first again to
teach about Bohr's atom. He had improved on Bohr's quantum work
by linking it to Einstein's relativity theory, and he, too, hoped eventually to understand quantum physics.

4

Independence and
Inner Freedom

On March 14, 1919, back in Berlin, one month after his divorce became official, Einstein suffered his fortieth birthday.
He did as little as he could to note the moment, but there was
no getting around the calendar. He had conceived a multitude of paradoxes about time and how it can slow down, but part of those paradoxes held that it is always the other fellow's time that passes more
slowly. Your own goes clippety-clop, clippety-clop. Adding to the
chariot's natural speed was the birthing of a new world, already half-
hatched in the four months since Armistice Day. At age 40, it was
natural to wonder, or at least for Einstein's friends to wonder, if he
would adapt to the changes or see them only as losses.

Einstein had appreciated especially the old world's internationalism. "A lost paradise," Einstein said of the time when ideas leapt between cities without anybody worrying about which nationality had
produced them. In 1911, Einstein had attended a groundbreaking conference financed by a Belgian, organized by Germans, chaired by a
Dutchman, and whose speakers included French and British scientists.
Travel between countries in those days meant simply boarding a train
or a ship and going. Only the most reactionary places, like Russia, had
sealed up their borders and demanded that foreigners show passports.

Now the mood was changing and borders had become as important to nations as skin is to its citizens.

March 1919 was a hard time to be in Berlin. The would-be revolution, suppressed in January, had returned with street-fighting and strikes. In an attempt to calm people the government splattered the walls with posters proclaiming *Socialism is here*. Too late, barricades and wires had sprung up all over the city, some erected by the revolutionaries, some by troops supporting the government. In Old Berlin's center, at the Alexander Platz, where an ugly statue of the spirit of Berlin rose four stories above the square, troops fired mortars at civilians who threw stones and packed pistols. The government declared a state of siege. Unlucky pedestrians were stopped by troops, directed to stand against the nearest wall, and executed on the spot. Rumor said Einstein had fled into hiding and was sleeping at a different place every night, though that story was false.

Amid all this confusion two Zionists approached Einstein. Many nationalist dreamers were in Paris that season lobbying at the peace conference, but Kurt Blumenfeld and Felix Rosenblueth had stayed behind in Berlin to continue recruiting prominent Jews to the Zionist demand for a Jewish nation. Einstein had known Zionists in Prague, where he had taught for two years at the German language university. He had not been impressed by the Zionist ambition that linked nationhood with religious identity, and the war had deepened his loathing for all nations. When the killing approached its height he wrote Lorentz in neutral Holland, "Men always need some idiotic fiction in the name of which they can face each other. Once it was religion. Now it is the state."

Lorentz shared the scorn for great-power ambitions, telling Einstein, "I am happy to belong to a nation which is too small to commit great stupidities." Einstein always loved Lorentz for that remark and was still quoting it many decades later. It was not an attitude likely to make him any kind of nationalist.

The Paris Peace Conference, which opened on January 8, aimed at restoring many internationalist ideals, shorn of the wrinkles that had led to years of murder. President Woodrow Wilson's list of 14 points opened with very much of an internationalist flavor and its last point called for a "general association of nations" to maintain the independence and territory of "great and small states alike." It seemed somehow both modern and faithful to the highest traditional ideals. But the war *had* happened and national sensitivities were tender to the touch.

Before the war people like Einstein had seen themselves as members of fellowships, societies with tastes and concerns common to a whole civilization. They were workers, or scientists, artists, or commercial men who each contributed to the growth and maintenance of a worldwide civilization. But in 1914 most of those people abandoned their general identities for national ones, proclaiming themselves French workers, British scientists, German writers, and so bloody on. Despite much urging, Einstein had resolutely held to an unqualified scientific identity. He was a physicist, not a German physicist. Early in the war he had been asked to sign a manifesto in defense of German culture, known outside Germany as "The Manifesto of the 93" or more commonly as "The Notorious Manifesto of the 93." He refused to sign. That was at Germany's zenith, when defying the pressure to sign looked like refusing to declare for the winning side. Now, with Germany in ruins, German nationalism was still not dead. Planck was again lobbying Einstein to choose sides and stay faithful to Germany in this difficult hour.

A question in the air for old-fashioned people like Einstein was whether fellowship itself was changed. During the war Einstein had refused to break with the fellowship of scientists. Now could he break with the fellowship of Germans? The starving Berliners whom Einstein found when he returned from well-fed, normal Switzerland were mere ghosts of past glory. Politically and personally desperate millions filled streets that before the war had been prosperous and abundant with optimism. Nighttime life in the former cosmopolis now offered seedy vaudevilles with third-rate singers and low-grade novelty acts like knife throwers, stage hypnotists, or the expert with a bullwhip who, with an artful crack of the wrist, could pluck a cigarette from a stage beauty's lips. On gray afternoons boys young enough to have been spared conscription kicked soccer balls across muddy fields. The slightly older lads and young men who had once been part of any street scene had gone for soldiers.

Of course, the prewar ideal that Einstein most cherished was the pursuit of knowledge, but he was getting on. By 40, most physicists have done their keenest work, although for the ultras among them this exhaustion is not always guaranteed. Across the channel in England, Ernest Rutherford had won the Nobel Prize before he turned 40 and then did his greatest work afterward—teasing out the atom's basic structure with its nucleus and orbiting electrons. In classical days Archimedes lived past 60 and appears to have done groundbreaking work all his life. Without his 60 years there would have been no medieval mathematics and the Renaissance would not have ended with Galileo's mathematical glories. Newton was 45 when he completed his *Principia* and laid the groundwork for the next two centuries of scientific coherence. Maxwell was 42 when he wrote his greatest book, the one that finally gave physicists some fundamental ideas that outpaced Newton.

And aged 40 or not, Einstein still had at least three things to learn. In logical order they began with finding evidence that would persuade his colleagues he was right about light quanta. He also had to figure out how light really worked with its wave-like v properties that somehow became a particle-like hv. Finally, he had to find a way to unite electromagnetic radiation and the rest of the cosmos under one general law. Einstein, being Einstein, devoted the most time to the third and largest task, linking electromagnetism and general relativity, rather than to assembling disciples who would join him in an effort to persuade physicists that light quanta were real.

Fourteen years after proposing light quanta, Einstein had only ever had one disciple on the matter. That was Johannes Stark, a lecturer at Göttingen. Stark was even less politically skilled than Einstein, for while Einstein simply paid no attention to winning support among his peers, Stark seemed to specialize in making enemies. In fact, historians of science sometimes suspect that Stark was attracted to Einstein's light quanta because the idea so dumbfounded other physicists. In 1909 Einstein had been asked to give his first public lecture. The continent's greatest physicists assembled in Mozart's city, Salzburg. Planck was there. Sommerfeld was present. Stark was on hand. Lorentz was there and so was Born. Many of Einstein's professional acquaintanceships began at that memorable meeting where he told his audience that physics must fuse waves and particles into one concept. After he spoke Planck rose to say he saw no reason yet to abandon Maxwell's theory and support light quanta. Stark, and Stark alone, came to Einstein's defense. That scene in Salzburg with Einstein and Stark standing by themselves in the midst of Europe's most honored physicists pretty well sums up Einstein's isolation. And then even Stark sat down.

For Stark and Einstein, light quanta had explanatory value, most notably in the photoelectric effect. Late in the nineteenth century Heinrich Hertz discovered that the presence of ultraviolet light increases an electrical discharge. This finding seemed as surprising as Ali Baba's discovery that saying "Open sesame" sent a boulder rolling aside; there was no reason to believe that the first phenomenon could in any way cause the second. Hertz's discovery preceded the discovery of electrons, so it was only in 1899 that J.J. Thomson proposed the modern view of the photoelectric effect: when light strikes metal it knocks out some electrons, the electrical discharge.

In 1905, Einstein accepted Thomson's idea that light might release electrons; however, to make the idea work, Einstein changed light's physical description. Thomson and others had assumed light was a wave, but Einstein proposed that light worked like the soccer balls that boys were always kicking about. He suggested that when a light ball hits an electron with enough energy, it sends the electron flying. And how much energy does the light ball carry? Why the frequency of the light (v) times h, of course. From that proposal, Einstein and Stark extended light quanta's role to other areas; however, by 1913 Stark had picked a quarrel with Einstein over who had first dibs on a quantum explanation for how light breaks up molecules, and Einstein lost his one light-quantum disciple.

Einstein was deeply admired. His approval was sought by colleagues who fairly danced when he praised them, yet he had no school, no followers who saw the same unresolved problems that he studied and who looked for explanations where he looked. Einstein had seen this pattern before. He drew followers only after a theory proved successful and by then he had moved on. He was more the solitary artist than the peer-group scientist. His work habits were less like, say, Ernest

Rutherford's than they were those of Thomas Mann, who that April resumed his self-appointed task of composing the story of *The Magic Mountain*. When you picture Einstein at work, imagine a figure in an isolated room hunched over a paper tablet pressing ideas onto paper. Do not envision anything like a laboratory where men in white coats work as a team to poke at nature.

That very spring a British team, in deep admiration for the theory of general relativity, was mounting an expedition to the South Atlantic to observe a solar eclipse and test one of relativity's predictions, but Einstein had been alone during the long years spent developing the theory. At the time he could not understand why every serious physicist in the world was not trying to repair the break he had demonstrated in Newton's logic. Before Einstein's theory of relativity, the concepts of motion and gravity formed a coherent theory. After Einstein, they did not. Surely that was important. Yet no one seemed to care. Planck loved Einstein's theory so much he had coined its name: relativity. Yet when Planck heard that Einstein was trying to revise the theory of gravity, he advised against it because failure was almost certain and success, if Einstein were granted that miracle, would not be believed. Yet Einstein had soldiered on—if one dares use that expression in his case—and found his success alone.

That had been during the prewar international fellowship, when schools of thought seemed to matter less. Einstein had been confident that success, contrary to Planck's prediction, would bring acceptance. He was never one like Stark to create factions among his fellows. He could have done so. In Berlin he directed the Kaiser Wilhelm Institute of Physics, a dummy organization created simply to bring Einstein to Berlin. Its office was wherever Einstein sat down, but it did have a small budget and Einstein did have grant money to distribute. He could have developed his office into anything, just as Walter Gropius did when that same spring of 1919 he opened his new Bauhaus school of design. Einstein could have used his grant money to support only research into his questions. Instead he sent money to theorists he thought worthy, no matter what their approach to physics. He never did become an intriguer.

Meanwhile, in Copenhagen, Niels Bohr was busy organizing his new Institute for Theoretical Physics as a training ground for young Bohrians. A few days before the Paris Peace Conference began, Bohr hired Betty Schultz to be his secretary. She would retain that post for more than 40 years, the rest of Bohr's life, and knowing Fru Schultz would become a sign that a physicist really had spent time in Copenhagen. Of course Gropius and Bohr were masters at founding schools and acquiring disciples, so it is impossible to argue that they succeeded only because of postwar changes. Nonetheless, both were very capable mammals lucky enough to appear just as the ruling dinosaurs exited stage left.

How shocked Einstein would have been to learn of postwar intrigues in science and art. Leon Brillouin, one of France's finest physicists and a man whom Einstein knew personally, actually preferred scientific factionalism and despised international learning. He wrote a letter that June protesting against inviting Germans and pro-German neutrals "whatever their scientific value" to international meetings. "I am thinking, for example of Debye, the Dutchman of great merit, who spent all the war as a professor in Göttingen. Naturally, also of Einstein who, whatever his genius, however great his antimilitarist sentiments, nonetheless spent the whole war in Berlin." No doubt Brillouin would have doubled that proposal if he knew that Einstein had recently written to Ehrenfest about Germany's continuing blockade, "Those countries whose victory during the war I had considered by far the lesser evil, I now consider only slightly less evil."

Planck did not know about Brillouin's letter either, but he sensed the victors' mood and he guessed at efforts to move Einstein into more acceptable settings. He persisted in urging Einstein to stay in Berlin. The Zionists, too, came back to urge him to think again about the Jewish nation. Blumenfeld and friend might not have felt their entreaties were hopeless because, unlike most anti-Zionist Jews, Einstein was not an assimilationist. He had no wish to become still more German. On the other hand, neither was he likely to see himself as part of a closed group, no matter how far the postwar world strayed from its prewar ideals. He still lived his life on his own terms, not because of events or loyalty, and he thought other people should live that way too. If he had not been a secular Jew, he would have made a fine radical Protestant. His moral opposition to telling others what to believe was as deep and decisive as it has been for the staunchest Anabaptist. From Einstein's perspective, men like Brillouin seemed to be taking revenge against themselves. He still held to the prewar assumption that silencing intelligent colleagues of whatever nationality risked reducing one's own ability to understand the world.

Planck was less of an oak and bent more with the winds. Before the war he had gone along with internationalism. When war erupted, he went along with patriotism and signed the notorious manifesto. When the war went badly he regretted having signed it. But one point remained stable in Planck's eyes—the greatness of Einstein and the enormous contribution he made to German physics by remaining in Berlin. Planck was among the first to recognize in Einstein's original relativity paper that a "new Copernicus" had appeared. Despite the name "relativity," both Einstein and Planck were interested in the absolutes of science. It was Planck's greatest pride that in the quantum's h he had discovered a physical constant, a number that never changed no matter what the physical context. The fear that postwar chaos might send Einstein from Germany struck Planck as too terrible a cost of defeat and he was determined never to pay it.

Einstein appreciated Planck's regard. He knew that Planck was not the most quick-witted scientist, but Einstein admired his doggedness. In October 1900 Planck had written down an equation that no experiment would defeat. Its predictions about radiation were exactly right and continued to be right as measurements grew more refined, but technical success was never enough to win Einstein's heart. What he admired was Planck's reaction to his own success. "On the very day when I formulated this [new radiation] law," Planck recalled, "I began to devote myself to the task of investing it with a real physical meaning." As it stood, Planck's formula used an abstract expression that Planck had just guessed at. It was an inspired stab in the dark that worked brilliantly, but why did it work? Planck immediately tried to understand what he had wrought.

That search for meaning won Einstein's heart. In the intense mental struggle that followed, Planck visualized radiant energy coming from an unknown number of vibrating atoms. His reasoning was mathematical and general, but Planck's ideas become clear enough if, for atoms, we imagine a group of springs, each one trembling for its own reasons, and each one sending its vibratory waves into space. Consider, for example, a set of box springs, maybe the one that supported Einstein's own mattress. Einstein himself can be on top making love to an actress. Real mattresses behave in the classical manner described by Newton, but Planck was trying to understand what peculiarities were needed to justify his radiation equation. So besides Einstein and the actress, we have a bed with imaginary quantum springs. As the couple begins to move, the springs stay still, but then the couple crosses some threshold of action and the springs begin to bounce, slowly and then with increasing vigor. The vibratory state of any spring is a bit uncertain. We just do not know enough about all those impromptu actions that are exciting the bedsprings, just as we cannot predict with certainty where a roulette ball will land. We know, however, that over the long run the casino owner and not the casino players will come out ahead, and we can also predict that the as bed's bouncing increases, the more energy the springs will emit. As a mathematical physicist Planck naturally designated the total energy by E.

Biologists and generalists will concentrate on the couple on the mattress, but physicists might find professional interest in the box springs below. Individually, each spring emits its own energy. Planck called the energy of an individual spring ϵ and used that letter to indicate the spring's energy, weighted to reflect the probability of it being in a particular vibratory state. The springs in the box's middle might be vibrating quite rapidly while those over toward the edge are almost still. Planck's theory had to take into account the probability of atoms (springs) shaking at different rates. The total energy available to the lovers will be the sum of all those trembling springs. Planck's challenge was to figure out how the energy from each of those springs added up to E. When probabilities enter the math, equations tend to grow more complicated, but, in essence, Plank's equation added every quantum's little ϵ.

Typically, mathematicians like to solve complex problems by shrinking parts of an equation down to 0 and then forgetting about that part, but Planck found that he could not shrink energy forever.

This is the point where normal bedsprings and our imaginary bedsprings differ. Normal bedsprings can always vibrate just a little less. Planck found his bedsprings cannot vibrate below some minimal amount of energy. That is, if a spring is to contribute any energy, it needs a minimal amount to get started and when it does start it moves with a pop as all the energy needed for that action is released as a unit. What's more, in these special springs, the energy continues to increase with little pops, so that the couple above, if they were not otherwise distracted, might notice that the bed's violence grows with leaps and bounds rather than with the steady, continuous growth of most things.

Planck's physical explanation for why his radiation equation worked was to show that it could be derived by assuming that a vibrator's radiating energy bundle combined some minimum chunk of action with the vibrator's frequency. That minimum action was Planck's beloved constant, h. The equation for the energy in each individual spring became $\epsilon = h\nu$. Planck did not think of $h\nu$ as a real thing the way Einstein did, but he had shown that by speaking in terms of units of action, what he called quanta, he could understand why his earlier radiation equation worked.

Einstein saw his own work as an effort to take the kind of step that Planck had managed, and he was frustrated by how little of quantum radiation made coherent physical sense. In a letter to Max Born that June Einstein said, "One really ought to be ashamed of [the quantum theory's] success because it has been obtained in accordance with the Jesuit maxim: 'Let not thy left hand know what thy right hand doeth.'" Even with the misattribution (the passage is from the Gospel of Saint Matthew) the joke flashes a glint of what the Einstein school (had there been such a thing) would have been about: finding physically meaningful coherence.

Mathematicians reign over a dreamlike world where they can insist that every rule and axiom fit logically and completely. Physicists, however, are stuck with untidy reality and they must often tolerate some contradiction between laws. They tend to view these discrepancies as unavoidable ragpiles to be cleared up later. Einstein wanted to make physics as simple and neat as mathematics and once asserted that "science by no means contents itself with formulating laws of experience. It seeks, on the contrary, to build up a logical system based on as few premises as possible, which contain all laws of nature as logical consequences." This distinction between laws of experience and laws of nature was basic to Einstein's thought. Laws of experience allow predictions but carry people no closer to deeper truths. They are merely efficient. Laws of nature are of another order and move people toward understanding how nature works.

Compared with the paradoxes and blank walls of the quantum, the postwar torments Germany faced with its vengeful conquerors offered Einstein a relaxing break. Right after moaning to Max Born about the Jesuitical quantum, he shifted to say, "I do not see the political situation as pessimistically as you. Conditions [in the Versailles treaty] are hard, but they will never be enforced." Meanwhile the rest of Germany was in agony over the Allies' insistence that Germany proclaim officially that the war had been all its fault.

The howls of distress could not distract Einstein for long, however. His thoughts always returned quickly to his science. In his ambition to create a perfectly logical, unified physics, Einstein had two heroes: Euclid and Newton. Euclid was the god of Einstein's logic; Newton the prophet of logical physics. Euclid had managed to derive the whole of plane geometry from a few axioms. Richard Feynman has pointed out that most of Euclid's theorems were known already to Babylonian and Egyptian engineers, but for them geometry's rules were technical facts that revealed no deeper reality. By showing their logical connections, Euclid transformed geometry from a mere set of rules into one of the pillars of wisdom. After Euclid, people recognized geometry's importance for all educated citizens. Philosophers and savants of all sorts turned to it as a source of understanding.

For thousands of years no other subject combined geometry's practical importance and philosophical meaning. Then Newton brought Euclid's transformation to physics, deriving his mechanics from a few laws of motion. Again, the practical side of this matter was already well known. Seventeenth century artillerymen and machine makers used the mechanical principles every day, but Newton's axioms became important laws of nature that all educated people needed to know if they were to think successfully about reality. Early in the twentieth century Einstein altered the foundations of Newtonian mechanics, but he was determined to retain a physics that mattered to ordinary people because it dug down to reality.

The unEinstein school rejected Einstein's great ambition and was satisfied to formulate those "laws of experience" that would predict accurately what would happen in particular cases. Successful laws of experience have been good enough to let civilizations invent their machinery, erect their buildings, and perform their experiments with confidence. The unEinsteinians doubted that people could discover laws that were so objective that they exposed the absolute reality behind our subjective experiences. Their science was one of technique and method with no extra implications that interested nonscientists.

During Einstein's schooldays common opinion taught that Euclid's geometry and Newton's physics were both true; yet there were skeptics. An Austrian physicist-philosopher-psychologist, Ernst Mach, argued that Newton's mechanics was philosophically incoherent and that it was sheer vanity to believe laws based on Newton's absolutes (space and time) described nature as it really is. At the same time, a French mathematician, Henri Poincaré, argued that Euclid's geometry was no more "true" than any of the other geometries that mathematicians had invented. Einstein studied both of these skeptics and, although he accepted their criticism of the established gods, he never joined Mach in believing that there were no real, natural laws to be had anywhere. Mach's criticism just made him think harder.

Mach did have his disciples, but they were a minority among physicists. The majority agreed with Planck, Einstein, and Lorentz that physical laws were more than laws of experience. If you looked outside physics, however, it was easy to find examples of scientific ideas that got at the experience of a thing, but plainly did not describe objective reality. The unreal, scientific doctrine causing the most commotion in 1919 came from Sigmund Freud. Although many people did take literally his notion of an unconscious, it was easy to accept that Freud's components of the unconscious—the ego, the id, and the superego—were useful fictions rather than physical realities lodged somewhere in the brain. An unEinstein physics would have held similarly that Einstein's laws of nature—the constancy of the speed of light, for example—were efficient concepts for working out what we experience but were not ultimate truths.

Before World War I, Einstein's view of science had been generally taken for granted, but by the war's end only very old-fashioned people, or, as in Einstein's and Planck's cases, people who had met unusually great success at reasoned efforts, could insist that the world made sense all the way down. In physics, the growing skepticism was heralded by 70-year-old Franz Exner, an experimental physicist of excellent reputation who had taught for decades in Vienna. In 1919 he published a lecture series in which he denied science's universal lawfulness. He suggested that even a falling body, if it were examined over short enough times, would be found to move at random, going up as often as down. Exner argued that nature's apparent lawfulness was merely a statistical illusion. Events that appear regular to our senses would, at the microscopic level, be random. Instead of digging down to the truth, this kind of scientist expected to dig down to lawless chaos.

The most eminent physicist in the unEinstein camp was Niels Bohr. He leaned Exner's way, suspecting that statistical randomness lay behind much experience. In the summer of 1919, Bohr wrote his old school chum, C.G. Darwin (Charles's grandson), that he was "inclined to take the most radical or mystical views imaginable" concerning quantum interactions between light and matter. By "most radical," Bohr meant he was willing to abandon the keystone of realistic science, the insistence that matter and energy is conserved. Conservation was the physicist's way of saying you cannot get something for nothing, or to use the proverb of a later age, there is no such thing as a free lunch. Without conservation, cause and effect would disappear. Law would disappear. We see a rock flying through the sky and know it must have come from somewhere. Rocks do not just appear out of nothing and nowhere. But they could if the conservation of matter did not hold. When something like a soccer ball begins to bounce across Berlin fields, the energy for the motion has to come from somewhere. Somebody or something sent it flying. Likewise, when a light quantum (if there were such a thing) passes through a camera's shutter and strikes the film, it must knock at least one atom about, the way a soccer ball knocks a goalie back, passing its energy into its target.

Bohr's letter to Darwin proposed that on the quantum level, this energy conservation did not apply. Therefore cause and effect did not apply. Meaningful law did not apply. He added that it was "quite out of the question" that photoelectricity had a causal explanation. Quantum effects did not begin with an energy input (that is, with a mechanical cause).

Darwin replied that Einstein had, years before, considered dropping the law of conservation, but found that quantum theory without energy conservation "was no better than with."

Einstein was ready to discuss these issues with anyone who looked him up. A good place to catch him was on the tram that he rode between the university and home. One who often chatted with him there was a philosophy student named Ilse Rosenthal-Schneider. She recalled how Einstein "was at all times ready to listen patiently to questions and to answer them in detail." He loved escaping what he called the "merely personal" in talks like these; yet the whirlpool of the personal pressed in. He closed his June letter to Max Born by saying: "With sincere regards to you and your wife, also from my wife." *My wife?* Yes, on June 2, 1919, Einstein and Elsa went to the registry office and quietly married. Berlin's intellectuals often sneered at Elsa and suggested she was unworthy of so splendid a thinker, but she served him as a valuable presence. She was good humored, proud of her genius husband, tolerant of his life in an imaginary world, and alert to his practical needs (such as reminding him to put on underwear). What she got in return—apart from status and income—is not so clear. She had had status and income before, but she divorced it, so perhaps she did indeed marry for love of her "Albertle."

The wedding had come only five days after the solar eclipse that would confirm or deny Einstein's great theory, but Elsa was not a scientist, professional or amateur. Her husband did talk to her about physics, but he was talkative by nature. Almost any sounding board would do, so long as the chosen ear was willing to listen to what Einstein said. Elsa joked that Einstein had explained relativity to her many times, but understanding it was "not necessary for my happiness."

Perhaps, however, being Einstein's wife was necessary for her happiness. Many people who wanted him as their champion did not easily give up when he showed no interest. The Zionists persisted throughout 1919's spring and summer, seeing Einstein's embrace as too valuable a prize to be abandoned without the utmost effort. Finally Kurt Blumenfeld saw his opening. Instead of talking about the Jews as a group, he drew Einstein's attention to nationalism's opposite side, its benefits for the individuals. Zionism was intended to "give Jews inner security . . . independence and inner freedom." Einstein believed everyone should have the strength and security to follow an internal compass, and he knew that anti-Semitism made that freedom more difficult. He began to take his Zionist missionaries more seriously. Meanwhile there was that fateful eclipse to wonder about. The British had taken their photographs and gone home, but what they saw remained unknown.

Like the Zionists, Planck would not give up. He pressed Einstein to make his commitment to stay in Berlin. Finally Einstein wrote to assure his friend that he would not abandon Germany, especially not as it was finally turning toward democracy. Germany had signed the peace treaty. The Allies had lifted the blockade. An optimist might think he had detected a dawn.

Hardly was the ink dry on his letter to Planck, however, than Einstein began to feel pressure from the Dutch to join them. Paul Ehrenfest wrote him to urge a move, promising that the normal maximum salary would be Einstein's minimum and that, "You can spend as much time as you want in Switzerland, or elsewhere, giving lectures, traveling, etc., provided only that one can say 'Einstein is in Leiden— in Leiden is Einstein'." Such offers would become so common in later decades that it would not seem odd for a university to boast about scholars without actually making them available to students, but it was very remarkable in 1919. Einstein, whose name was still unknown to most of the world, had become in certain circles like a champion athlete whose endorsement is sought and readily paid for, but whose actual contribution to the endorsee is a matter of indifference.

And the Zionists were still pitching woo. Anti-Semitism was beginning to swell in Germany. Impotent resentment against their conquerors' demands was being redirected against the Jews, the way a frightened, wounded elephant, afraid to charge a hunter will rip apart a thorn bush in misdirected fury. Eric Warburg reported seeing Einstein give a lecture on relativity and hearing students shout anti-Semitic slurs as he spoke. In the end, Einstein told his Zionist pursuers, "I am against nationalism, but for the Jewish cause." That remark and the letter of reassurance to Planck were probably the first indicators that the internationalist Einstein would find some place in the new world of nations and national factions. As for the postwar taste for ignorance and its skepticism of reason, no, there were never such indicators.

5

A Mercy of Fate

On Saturday evening, October 25, 1919, Albert Einstein walked alongside a small canal toward a palace called the Trippenhuis. He was in Amsterdam, which styles itself as the Venice of the North, but he found no Venetian palazzo. Instead of being light and graceful, the Trippenhuis is as heavy as a prison. If Einstein paused to study its weighty facade, he saw olive branches surrounding cannon and other weapons. The palace's builders had grown rich supplying arms for the Thirty Years War. Once through the Trippenhuis doors Einstein found a more ordinary interior with large rooms and painted ceilings. The palace had been taken over by the Royal Dutch Academy of Arts and Sciences. Einstein had come not to a ball, but to a science gathering.

The bright lights that greeted him shone on a room filled with physicists. The Netherlands, despite its small size, was home to many theoretical physicists and most had turned out for the opportunity to view the great Einstein on his triumphal night. He saw that Lorentz was there and Ehrenfest had come up from Leiden. Lorentz and Ehrenfest together made a Dutch Mutt and Jeff team. Lorentz was tall, lean, always photographed wearing a black bow tie. Ehrenfest was short, square-shaped, less formal in appearance. Besides meeting the star pair, Einstein was also greeted by less-remembered figures like 1913's Nobel Prize winner, Helke Kamerlingh-Onnes, the discoverer of superconductivity at very low temperatures. The scientists buzzed about the meeting room, talking of serious and temporary things. Einstein was not yet a figure of myth, and as he greeted smiling faces there were no representatives of the Dutch press on hand. They were not invited, probably because the evening was not entirely in line with scientific etiquette. Lorentz, the grand old man of Dutch physics, the man Einstein admired most in the world, was going to report some data (the data of the century, as it turned out) before it was announced by the team that had gathered it.

Ehrenfest was as proud as a happy brother. He admired Einstein so much he had once considered moving his family to Zürich, even with no job promise, just to be close to his hero. Einstein moved among these thinkers with an alert confidence. In autumn 1919 he looked better, less worn, than he had a year earlier, and when a man is the hero of the hour, meeting in a 250-year-old setting, he looks still better. The evening's conversations went unrecorded, but presumably Einstein was greeted enthusiastically and given many anticipatory congratulations.

Ehrenfest was excited by a visit that Niels Bohr had paid recently to Leiden. He had been deeply impressed by Bohr's lecture. Presumably, as Einstein and Ehrenfest chatted, friends and bold strangers came up to note that it was a grand night. Ehrenfest and his wife, Tatiana, were experts in the statistics of thermodynamics. Together they had written a book that made the field intelligible to their fellow physicists. Einstein much admired them and their brains. Paul was a complicated man who combined his many talents with hero worship and bouts of depression. He doubted his own abilities and would write mournful letters to the hero-in-chief. Late in the war, Einstein had scolded him, "Stupid you certainly are not, except insofar as you keep thinking about whether or not you are stupid. So away with the hypochondria! Rejoice with your family in the beautiful land of life."

As for Ehrenfest's enthusiasm for Bohr, Einstein did not yet share it. He had never met the Dane and did not quite catch what it was that made him so well admired. Bohr's work sprang from none of physics' usual techniques. Instead of deducing ideas from known laws, Bohr argued from blatant assertion and did not show the math that supported his conclusion. It was hard for mortals to know just where

Bohr found his insights, but he was rich in them. Ehrenfest spoke so insistently that Einstein reported back to Planck, "[Bohr's] must be a first rate mind, extremely critical and farseeing, which never loses track of the grand design."

Einstein had some Berlin gossip of his own. Planck's life had become a terrible heartache. First his son lost in the war, then his daughter in childbirth, but now a second daughter, the twin sister of the first, was pregnant, so there might be some happiness soon.

Surely, as the evening persisted, the air filled with tobacco's sweet taste and surely, too, there were still more preliminary congratulations for Einstein. The Dutch physicists had assembled to hear the result of the British expedition's test of Einstein's theory. Once, Einstein probably would have gone to London and joined the expedition's leaders in announcing their results; however, the war had transformed the Netherlands into a neutral corner for exchanging information between enemy powers. When the war was at its height, news of Einstein's general theory of relativity had skirted the western front and reached Britain via Holland. In London, ideas from Germany were viewed as greetings from Lucifer, but the astronomer Arthur Eddington was a Quaker and a pacifist, so he peeked. The theory piqued back. Eddington went to the trouble of mastering Einstein's difficult ideas. When the war ended he organized an expedition to see if a coming solar eclipse matched Einstein's theory. The team's photographic analysis took much longer than anticipated. The silence persisted for weeks beyond expectations. At first Einstein waited patiently, showing no concern until, unable to sit still, he asked Lorentz if he had heard anything. Lorentz contacted England and eventually sent Einstein a wonderfully ambiguous telegram: "Eddington found star displacement at rim of sun, preliminary measurements between nine-tenths of a second and twice that value." This confirmed, as Einstein had expected, that light bends when it passes through a gravitational field, just as it bends when it passes through a lens. But "nine-tenths of a second" was the bend that Newton's old measure for gravity would have supported while "twice that" was the prediction made by Einstein's measure. Now Lorentz had newer, more decisive information to announce.

The Dutch setting that night could have been Einstein's to keep—

learned minds, a sophisticated city, a neutral country sure to be spared the many agonies that were obviously Germany's immediate destiny, but with his usual No–Yes Einstein had turned down Ehrenfest's and Leiden's offer to settle there. No, he would not move to Leiden; yes, he would come to give regular lectures. He had tied his fate to German democracy's, and so, instead of appearing that night as Holland's latest star, he held his more accustomed role of honored outsider, a triple outsider.

Einstein's friend Walther Rathenau told an acquaintance what it was like those days to be such an outsider. Foreigners, the acquaintance remembered, "are unanimous in their attitude of condolence that he should belong to [the German] nation which they regard with a mixture of loathing and contempt unique in history. Their behavior is the same as Christians adopt towards outstanding Jews whom they accept but pity because of their awful Jewish connections. As a Jew, he is perfectly familiar with such politely disdainful turns of phrase and the accompanying looks. What seems harsh to him is that, having had to put up with this all his life, he is called on to endure it a second time as a German."

Einstein, however, loved the freedom that came from being such a universal outsider. Yet he was shrewd enough to know that the outsider's success depends on the insider's recognition of special merit. He was in because he was so often right. If he were ever wrong, he could not count on the old boys' club to look out for him. So he had more than abstract theory riding on the data from the Eddington expedition. He had bet his fate on the proposition that nature is, in its roots, profoundly simple, that the *why* of it can be fully grasped in accordance with a few basic natural laws.

In future years Einstein would be famous for being difficult to understand, but his difficulty arose from his simplicity. Others would find it hard to throw away complicating ideas, just as would-be suitors find it hard to shed the extra weights that keep them from floating easily on a sea of love. Years before, when Einstein was an unknown patent officer, there were several insiders who almost hit on relativity. Lorentz had worked out the mathematics, but he kept in a complicating idea—ether, the mysterious substance said to fill the cosmos and provide an absolute reference point for any event in the universe. France's star mathematician, Henri Poincaré, also came close to relativity, but he was never able to whittle his system down, as Einstein did, to two basic axioms. Poincaré's too-complicated theory left him like a bloodhound who is subtle enough to sniff the right ground, but not quite good enough to find the prize.

Under the chandeliers, amid the tobacco cloud, the physicists took their seats. Lorentz stood before them, handsome and alert in old age. Moments like these always begin with preliminaries and asides, but he came to the point. He read aloud Eddington's measurements and explained their meaning. Starlight passing the sun had bent 1.8 seconds from its normal course. The number was exactly the one general relativity had predicted. Einstein had won; the age of Newton was over.

"It is a mercy of fate that I was allowed to live to see this," Einstein wrote Planck as soon as he knew Eddington's results.

The next day the Dutch newspapers were silent.

Einstein did not expect publicity or much public interest, although it was evident that there would be some stir when the news broke officially. Rewriting the law of gravity is no small thing, even for café chatter, but ultimately people want to know *So what?* In relativity's case, Einstein had no answer. In September, he had sent a letter to Max Born's wife discussing the limits of science, "the causal way of looking at things ... always answers only the question 'Why?' but never the question 'To what end?' No utility principle and no natural selection will make us get over that." Relativity had nothing to say about life's purposes, so Einstein was unprepared for the way his world tipped over on November 7 and stayed tipped over. The London *Times* that morning published a report on the preceding afternoon's joint meeting of the Royal Society and Royal Astronomical Society, where, beneath a portrait of Isaac Newton, Eddington reported and discussed his expedition's results. The *Times* story attracted notice and throughout the day reporters were calling and visiting Einstein's apartment. There had been other physicist celebrities like Roentgen and Curie, but even so there was something unprecedented about the depth and persistence of the Einstein fascination. In a morning he became a symbol of science, modernity, and genius under a shag of crazy hair.

Eight days later the Nobel committee announced its physics prize. Max Planck was awarded, a year late, the prize for 1918. The 1919 prize went to Einstein's old disciple, Johannes Stark. Normally a physicist winning the Nobel gets to be, for a day at least, the world's most esteemed scientist, but with all the world shouting and writing about Einstein, very few crumbs were left for the Nobelists. What crumbs there were went more for Planck. With Stark ignored even at the promised pinnacle, Einstein's fame had produced a jealous, relentless enemy.

6

Picturesque Phrases

lthough it was springtime, Niels Bohr arrived at the Einstein
apartment in the guise and spirit of Father Christmas. He
carried gifts and good will from Denmark's untroubled countryside, a land overflowing with butter, cheeses, hams, the splendid fats
and sugars of a wealthy world. The gifts were rendered almost tragic
when spread on the spare table of Berlin's most celebrated man. The
Einstein household was a particularly sad place that April. In March
Einstein's mother had died miserably in the apartment, suffering from
stomach cancer. Einstein had seemed to many people, including himself, as a distant brain, unattached emotionally to anyone in the world,
but he wept when his mother died. It is hard to suppose that any man
as confident as Einstein did not enjoy a beaming mother's presence
when it counted. And even in the months just before her death,
Einstein liked to brag to her. When he heard the first hint that the
Eddington expedition supported his theory, he immediately sent a
note to his mother boasting that he was proven right. Bohr's smorgas-
bord from "Neutralia" (as Einstein called it) was an unexpected pleasure in a house of mourning.

Both Einstein and Bohr claimed to have been instantly attracted
to the other. In part, this mutual admiration was the inevitable result of
their work, but they also had personal reasons to take a liking. In

particular, neither respected pomp and titles. When Bohr became a professor he was, as custom demanded, presented to his majesty, Christian X, king of Denmark. Later the king would be widely admired for his refusal to accept the anti-Jewish legislation of his Nazi occupiers, and even before his meeting with Bohr the king had advanced Denmark beyond the United States by giving women the vote. Still, he was a formal man with a military bearing and no instinct for spontaneity. Bohr, dressed in morning coat, was an informal man who had done well to remember to keep his white gloves on when he shook the king's hand. His majesty told Bohr he was pleased to meet Denmark's great soccer player. The correct response was, "Thank you, sire," followed by a step back. Bohr, however, replied along the lines of, *Thanks, king, but it is my brother who was the soccer star.* Christian X was startled and tried again. He said once more how pleased he was to meet Denmark's great soccer player, and Bohr once again corrected his monarch. With that second contradiction, the king abruptly announced, "The audience is over," and it was. No wonder Einstein felt a natural attraction to his Danish guest.

Elsa was delighted by the food package and its simple pleasures. Berlin was in a tangle. A transit strike had forced Einstein to guide Bohr on foot to his apartment. The two men hiked for miles from Planck's home, where Bohr was staying in suburban Dahlem. Einstein showed the way through Berlin's southwestern district. That's a wonderful scene to imagine—two absent-minded dreamers feeling their way through the Prussian metropolis while lost in talk of physics. They could have wandered anywhere without noticing and suddenly discovered they had no idea where they had landed or how to find a route back. Those two could not count on boarding a tram without getting lost. Luckily for them, Dahlem's main route, the Archiv-Strasse, was wide and easy to follow. They could track its curving miles and cross its squares without looking up. Bird-stained statues occasionally looked down on them, but they probably did not return the stare.

Planck's sufferings offered a deep well for sympathetic gossip. Bohr's host was 62 and in good health, but he still seemed to prove that old age is a curse. His second daughter, Emma, had died in childbirth, just as her twin sister had. "Planck's misfortune wrings my heart,"

Einstein told Max Born, "I could not hold back tears when I saw him."

Both Bohr and Einstein were theorists rather than experimentalists, although Bohr's work was more guided by experimental results. He depended on lab work to refine his theories. His picture of the atom had been very simple at first, but changes in observed magnetic and electric fields led to many complications. For him theory was a way of describing and anticipating experimental data.

Einstein's ideas did not work like that. He laid a field's ground rules and then let others worry about loose ends, although somehow with Einstein's work loose ends were as rare as middle-aged soccer stars.

Like many people talking and disputing that day in Berlin, they disagreed profoundly on where to look for authority. Einstein preferred logic, Bohr leaned toward analogies. In these tastes neither seemed like the stereotypical, fact-bound scientist. Of course, scientists, like mathematicians, do appreciate logic and some, though hardly all, share a taste for analogies. Only gradually over the coming years would the implications of the difference between Einstein's mathematical bent and Bohr's more verbal leanings become apparent.

Bohr was a bit younger than Einstein, six years, and had an even more junior presence. He was shorter, but more gangly, with a long face and thick hair. The pressing issue for him was making sense of atoms and the light they emitted. In his Berlin lecture, Bohr coined the name "correspondence principle" for a technique that allowed him to use classical ideas to solve quantum problems. This rule of thumb let Bohr say, in effect, we have no idea how to solve this puzzle according to quantum rules, but probably older, non-quantum theory is not too far off. Following his analogy that quanta matches classical, he would use old-fashioned calculation methods and come up with answers supported by experiment.

Einstein appreciated the principle's value. When trying to grasp radically new ideas, old notions provide a stepping stone. His own theories of relativity had shown that in most cases Newton's old law of gravity works fine and that Newton's mechanics give wonderfully accurate results at velocities that keep well below the speed of light.

Newton was no longer useful in describing how the universe really works, but in most cases his ideas remain invaluable in calculating what will happen in any particular case. So if Bohr found a way to use well-understood ideas to glimpse what was going on in otherwise badly understood cases, more glory to him.

Most physicists could not understand Bohr's correspondence principle. Arnold Sommerfeld complained that it worked like a "magic wand"; wave it and Bohr got results that worked but made no sense. The wand's success made Bohr himself seem like a sorcerer. Try this, he would say, and it worked. Einstein, however, did not judge ideas solely by their success. He was no pragmatist, either in life or in science. He wanted to know *why* something worked. For him the correspondence principle's success was a clue, not a solution. The idea that something worked but had no physical meaning told Einstein that the idea was not yet complete. Yet Bohr seemed content to plow forward with his analogy. It was when analogies no longer worked that he became unhappy.

Eventually the two men reached Wilmersdorf, a fast-growing suburb at Berlin's southern edge. The population was changing so surprisingly that the village would soon build a mosque. The twists and turns from street to street would have demanded some attention, but the two talkers managed to make their way through it.

The radiation mystery that held both men's attention at the time focused on the light emitted by individual atoms. Just as a steel furnace gives off light, so does a single atom when sufficiently heated. Atoms are not miniature suns, giving off light in all colors, visible and invisible. They emit only a few colors which, when passed through a prism, appear as lines of color instead of a complete rainbow. Atoms also absorb light of specific frequencies, so when sunlight passes through a cloud formed by some element or another, the cloud's atoms remove some of the colors from the light, making black lines appear in the rainbow. Spectral lines can serve as the atom's fingerprints or DNA markings. They allow an investigator to know which atoms are present in something too hot or too distant to touch. It is the spectral lines that tell astronomers which elements are found in the stars.

Bohr's work concentrated on interpreting the atomic spectrum.

Einstein was interested in the spectrum's light—its light quanta. Trotting along, reaching clumsily for words, Bohr could be desperately inarticulate, but he still made it clear that he did not believe in Einstein's light quanta. Einstein, meanwhile, could drive his words straight to the heart of the matter, and he used a dry wit to make his case. Years later Bohr recalled Einstein that day using "such picturesque phrases as 'ghost waves guiding the photons'." Unfortunately for historical precision, Einstein could not have said exactly that. The term "photon" would not replace "light quanta" in anybody's vocabulary for many years yet, and Einstein was slow to adopt it even after the word was introduced.

The main route from Wilmersdorf into Berlin was the Kaiser Allee, one of those wide, handsome boulevards that allow eager talkers to move along steadily while ignoring their dignified surroundings. Even so, Einstein would have had to be awake enough to spot their turn onto the narrower, undistinguished Gunzel Strasse.

Except for the word "photon," Bohr's recollection of ghost waves makes sense. At that time Einstein was pondering ghost waves or ghost fields that would steer the light quanta. The ghost-wave notion hinted at ways to link particles and waves. A particle might ride a wave like a master surfboarder. In physics, the closest thing to a union of wave and particle is known as a wave packet. Imagine a bull whip in the hands of a Berlin cabaret performer, a man dressed in tight, shiny clothes who is skilled enough to snap the cigarette from his lovely assistant's mouth. Is that whip a thing of matter or is it energy? For the performer it is matter. He controls it, moving it in his hand like any other piece of solid earth. For the assistant, however, it is pure energy, exploding the cigarette from her lips with a poof. And if, God forbid, the performer slips and the whip kisses his assistant on the cheek, her face will explode in pain and her flesh will dissolve in an instant, leaving a hole where the sudden tide of energy rushed in. Meanwhile, the audience is aroused and confused. It does not know what it saw. The whip began as matter, but as it flew it disappeared into a blur and ended with energy's crack. So it is left to the scientist to unravel the incident. Filming the whole thing in superslow motion to capture the snap in one thousand frames per second, and looking at the footage, the observer sees something rare. Yes, the whip is solid all right, part of physical matter, but moving along the whip is a single bump of energy. It travels through the whip's length, twisting and distorting the matter as it goes. That bump is a wave packet, matter and energy, united into a single thing. The bump is as close as a wave ever gets to being a unit of something, moving as a single form through the whip, sticking together, going one place and not another, knocking the cigarette from the assistant's lip as surely as if it had been hit by a bullet. And yet it is clearly a wave moving through the whip, carrying energy as it flows. That wave packet is duality made briefly visible, displaying the properties of unit and wave together.

Einstein's street, Haberland, was a single block with modern (for 1920) residences, apartments rather than townhouses. Einstein and Bohr turned into number 5, and passed the doorman. Bohr greeted Elsa with his Father Christmas smile, producing the Danish foods that he and his elfin helper had hauled across miles of greater Berlin. Einstein's home was a large comfortable space, rather like the slice of Berlin that the two physicists had just crossed. Big and bourgeois, it provided a safe stage that Einstein could use and ignore at the same time. There was an ample dining room where Elsa, Bohr, and Einstein could sit and talk amidst wine and delicacies. Both men enjoyed pipes whenever they could and Einstein's apartment was a free-smoking-zone. That cloud and the inevitable chatter about physics would have eventually chased Elsa from the scene.

Bohr was not converted to light quanta by Einstein's ghost fields, and Einstein was not yet ready to convert him. The ghost-wave idea still had many problems. Especially, one could ask, what in nature serves as the cabaret performer's whip and carries the wave packet? In pre-Einstein days people said that light waves moved through something called the ether, but Einstein had shown there was no ether. The ghost wave had as yet no physical meaning, and until he could take the step from grasping the idea's utility to understanding how it worked, Einstein was unlikely to insist on his thoughts.

Bohr's objectives were different. He wanted to understand light in concrete language and proven analogies. In particular, he did not want to contradict the correspondence principle. Powerful as his magic wand had proved, however, it had nothing to say about light quanta.

A radical like Einstein was perplexed by Bohr's distaste for the anti-classical. Einstein was always ready to change ideas. He had broken with Newtonian concepts of absolute time and space when he proposed his theory of relativity. Then he modified major ideas in that first theory when he developed his second (general) theory of relativity. He knew he would have to revise those ideas still again if he ever succeeded at unifying relativity and electricity. So he was hardly likely to pause because some traditional analogy could not handle the new facts of the matter.

Bohr's preference for working with established concepts matched the way most physicists worked. Scientists typically are not so ready to trade in old ideas for the latest model. Despite relativity, many physicists, including Einstein's beloved Lorentz, still believed in an ether that carried light through the universe the way air carries sound from mouth to ear. Most scientists were on Bohr's side in this light quanta dispute.

This quarrel over light had a long history. Newton's light experiments sent a beam through complex arrangements of prisms to show that light moved in straight lines. When a prism appeared to bend a light ray, Newton said the light behaved just like a tennis ball striking an oblique racket. Newton's great English rival had been Robert Hooke. They quarreled constantly; however, Newton outlived his rival and had the pleasure of personally ordering the removal of the late Mr. Hooke's portrait from the Royal Society wall. From the start, Hooke had disputed Newton's particle theory by insisting that light was a wave that could be split in two. Bohr and Einstein, running on like rivers in springtime, were continuing a very old dispute.

For Einstein, a man like Bohr, who was devoted to explaining the world in visualizable terms was bound to be attractive; yet the two men's compasses pointed toward different poles. Bohr was a pacer, an active man who radiated energy, circling like a predator looking for an opportunity to dash in and catch something live. Einstein was more of a still point, a lamp that shed light on whatever came near him. And they disagreed about physics' heart. Bohr had come to Berlin to give a lecture. In it he publicly refused to quarrel with Einstein, saying simply, "I shall not here discuss the familiar difficulties to which the 'hypothesis of light quanta' leads . . . I shall not consider the problem of the nature of radiation," but in his visit with the Einsteins the nature of radiation is exactly what he discussed.

Bohr came at physics with very different tools from his host's. Einstein sat taking pleasure in the delicacies that his colleague had brought. He wanted very badly to take one more step and understand radiation. Bohr, the pacer, did not look for meaningful steps and had made his discoveries without them. Einstein, even when surrounded by Bohr's savory gifts, was apt to forget food when he became engrossed in talk. He once ate a pot of caviar without noticing that he was tasting something unusual. Still, the image of Einstein sitting in an overstuffed chair, using picturesque language, while Bohr paced about stringing together clumsy phrases probably gets close to what happened that day.

Strangely, Albert Einstein, the man living in the capital of physics, the conqueror of Isaac Newton, the world's most acclaimed scientist, was an outsider, while Niels Bohr, the lone physicist from remotest Copenhagen, was the insider. Bohr had studied at Manchester with Rutherford, the greatest experimentalist of that day, and corresponded regularly with him after returning to Denmark. Rutherford had good reason to appreciate Bohr as the man who had firmly established the theory of the empty atom. Rutherford's atomic theory had not been welcomed at first because philosophers had for thousands of years insisted that atoms, if they existed at all, were solid and indivisible, like a tiny piece of sand. A coal lump was a chunk of smaller, solid bits of coal. Rutherford had proposed, on the basis of his experiments, that atoms were mostly empty. Coal, he said, consists chiefly of empty space lightly sprinkled with electrons orbiting atomic nuclei. Spacious atoms made a radical concept and many physicists were uncertain about them, but Bohr was a loyal Rutherfordian and used the empty atom in his own quantum model. So Bohr's success became Rutherford's success, too, and won Bohr a prominent ally.

Einstein's work did not bring him such automatic constituents. He did not build on the theories of others. The one great exception had been his attention to Planck's quantum, yet even there he alarmed his potential ally by stressing the idea's revolutionary side. Planck was no radical and he spent many years looking to see if there was some way he could derevolutionize his quantum. Meanwhile, Germany's leading experimentalist and the man whose work provided the original evidence for light quanta was Philipp Lenard, the Nobel Prize winner for 1905. Unfortunately, he detested Einstein's approach to physics, so there was no Einstein-Lenard bloc the way there was a united Bohr-Rutherford front.

Bohr also attracted allies by generating important work for them to do. He had leapt onto physics' center stage in 1913 when he published a series of three papers that described the hydrogen atom in quantum terms. His basic idea was beautifully simple. In the hydrogen atom one electron orbits the atom's nucleus, just as the moon orbits the earth. From time to time—and this is unlike anything our moon does—the electron changes its orbit. When it does so it absorbs or emits energy. How much energy? A quantum's worth. Other physicists could sink their teeth into Bohr's idea. They could apply the theory to more complicated atoms, or look at what happens in magnetic and electric fields. All this work showed that Bohr's theory was powerful, yet incomplete. More work had to be done and physicists in Cambridge, Munich, and Göttingen rose to the challenge.

Einstein's work was different. He was like Moses come down from the mountain with his laws written on tablets. You could accept them, as happened with relativity, and apply them as needed, or you could refuse them and continue to worship Baal or some other idol. In neither case did schools pop up to sort through any loose ends. Either there were no loose ends, or they were so loose that only an adventurer with Einstein's abilities and obsessions dared confront them.

Bohr's more pragmatic approach did not appeal to Einstein. When first asked what he thought of Bohr's atom, Einstein had replied with a polite dismissal, "Important, if right." That was the comment he sometimes made when asked about a notion that made no physical sense and had no evidence to support it. Bohr's theory required the electron to change its orbit instantly. That is to say, the electron is following one path and suddenly is someplace else following a different path. This leap from one place to another without going through the space between—the so-called *quantum jump*—made no more physical sense than the Virgin's assumption into heaven. Einstein's calling it "important if" was typically sly.

Einstein abandoned his scorn for Bohr's atom, however, when he learned that the Bohr model could explain a seeming contradiction in Bohr's own theory. The spectral lines of some hydrogen atoms did not match the predicted pattern. Bohr showed that his equation worked perfectly if you supposed the atom was really a helium atom with a helium nucleus but only one electron. Einstein was stunned by this bonus success and changed his attitude right away, "This is an enormous achievement. The theory of Bohr must then be right."

Scientists often link two seemingly unrelated facts as Bohr had linked atomic structure with its spectral lines, but scientists usually make the link by finding an unexpected common ground. Newton had found a law that linked the way apples fall to earth with the way the moon orbits the earth. But Bohr made his link without adding such an explanatory idea. His equation used the quantum $h\nu$ but the quantum had no meaning in the equation. It explained nothing. It just was. Yet it worked and worked with astounding precision. Even years later, Einstein expressed his wonder at Bohr's success, writing, "That [quantum radiation's] insecure and contradictory foundation was sufficient to enable a man of Bohr's unique instinct and tact to discover the major laws of the spectral lines and of the electron-shells of the atoms together with their significance for chemistry appeared to me like a miracle—and appears to me as a miracle even today. This is the highest form of musicality in the sphere of thought." No wonder then that Paul Ehrenfest had reacted to Bohr's success by saying, "If this is the goal, I must give up doing physics."

Einstein and Bohr's first meeting had something of the mutual caution you expect whenever a prophet and a poet come together. Each knew the other was doing something related to and yet profoundly unlike his own work. It produced an Alphonse and Gaston exaggeration to their politeness.—*After you, Herr Bohr.*—*No, after you, Herr Professor Einstein.* And their mutual thank you letters have the tone of the obsequious headwaiter who has received an overlarge tip.—Einstein: "Not often in life has a person, by his mere presence, given me such joy as you did."—Bohr: "To me it was one of the greatest pleasures ever to meet you and talk with you."

Einstein flashed his cards a bit later in his thank you note. "I have learned much from you, especially about your attitude toward scientific matters." *Attitude*, not facts or theory. Bohr's poetic attitude, his radical pragmatism, his search for analogies that worked without deepening physical meaning was as far from Einstein's realistic attitude as a scientific approach could be. During the coming years their well-mannered dispute would strip scientific imagination naked, showing plainly what it strives for, what authority it recognizes, and what part is a matter of personal taste.

7

Scientific Dada

The wheels and whirligigs on the worldwide fame machine continued to spin out Albert Einstein's name. Articles, books, and photographs spilled from the presses and people studied them with care. His strong features and free-spirited hair were shown around the world. It was as though somebody had given him another one of those ad hoc chits, like the one Ebert had handed him at the Reich Chancellery, only this one said he had authority for absolutely anything. The name "Einstein" or the word "relativity" were frequently overheard while strolling past Berlin's Brandenburg Gate, or while tasting ale in a London pub, or huddling on a seat in a crowded New York bus. Einstein found the transformation of his name and face into a public commodity both astonishing and distracting, but the change revealed something unexpected about Einstein. He was good at being a star. He did not hire a publicist nor did he make endorsements, but neither did he flee the spotlight when it lit on him. He presented himself as genius surprised by acclaim, and the world loved him for it.

Many of fame's fruits resembled things he had known before, only now their creation had become industrialized. Before growing famous, he had received occasional invitations to speak in places where he was known, like Zürich and Leiden. Now requests to speak flooded him from places where he knew nobody. Before, he had received mail from strangers who knew his physics and had an idea to promote. Now many people who knew no physics sent him letters that sought approval for every kind of idea, or asked his advice about a universe of unscientific things. *For whom should I vote? Is there a God?* And not only did they write, they came to his front door. Elsa became her husband's butler and doorkeeper, letting a few through and sending many away. People recognized him on the street and said Hello. Before, he posed from time to time for a photograph with a friend; now strangers wanted to stand beside him while he smiled generously and a third person snapped a picture.

It was absurd and yet it was not quite ridiculous. Behind the frenzy and the comedy was a deep interest in science and its secrets. Every educated person in the world knew that Newton and Euclid had been reason's prophets. Suddenly the papers reported that they were overthrown; naturally people wanted to know what the new prophet had to say. During his obscurity a German publisher had commissioned Einstein to write a short book on his theory of relativity. Now that book was translated into many languages and a host of other writers produced books about relativity—more than 100 new ones in Einstein's first year of fame. Einstein had been embraced as the public's champion on a quest to make sense of the world.

Germany's scientific community was startled by the acclaim given its most ingenious member. Popular uproars over science were not to be trusted and were often about nonsense like some charlatan's claim to having invented a perpetual motion machine. Relativity was not nonsense and Einstein was no charlatan, but his ideas were expressed in terribly difficult equations. His ideas were becoming garbled and trivialized in this riot of fame. The mania's superficiality was amply illustrated by its narrow focus on relativity. Einstein's other work, most notably his work on quantum radiation, was ignored.

There were dissenters from the Einstein mania: scientists who did not like relativity, dogmatists who did not like science, and foreigners who did not like Germans. Among Germans there was a sudden pride that their ruins had yielded a world-honored thinker, but an especially strong anti-relativity, anti-Einstein movement also appeared in Germany. Einstein's Jewish blood, his anti-militarist history, his democratic principles, and his popularity in enemy nations made him a hated target of German nationalists.

An engineer of no reputation named Paul Weyland burst onto Germany's stage by denouncing Einstein's theory as "scientific dada." Part of the fame machine turned Weyland's way and he did what he could to hold the spotlight. He organized anti-relativity talks all over Germany. Weyland himself was to be the star orator at the Berlin meeting, for which he hired the Philharmonic Hall where Richard Strauss and Gustav Mahler once conducted. Bitter anti-Semites occupied the hall's opulent interior, boxes, and balconies. Already some extremists wore the swastika as badges and lapel pins. There were also scientists in the crowd who had come anonymously for a peek.

Max von Laue, one of Einstein's earliest admirers, was there. Fifteen years earlier Laue had been so startled by Einstein's brilliance in the first theory of relativity that he traveled to Switzerland to meet the unknown theorist. Laue expected to find an academic and was surprised to find a low-level civil servant, a clerk in a patent office, a man who spent his working hours reading descriptions of electric machines and how they were supposed to work. Einstein needed only a quarter of his wits for that job and was happy to entertain the young emissary from Berlin with the other three-quarters. The two men became immediate friends. They began a regular correspondence that would last even after Hitler kidnapped the German soul and Laue persisted as a rare, loud voice of Berlin science insisting that Einstein was a glory of German history.

The blatant anti-Semitic leaflets and signs at the entrance to the philharmonic startled Laue and he was shocked still again by Weyland's fiery speech. The master of ceremonies, Laue reported, could "compete with the most unscrupulous demagogue," although Weyland's speech could at least keep its audience awake. The evening also included more academically accredited speakers who served their purpose by giving shorter, boring speeches whose very tiresomeness gave them authority's aura.

Einstein had his own talents when it came to holding the spotlight. Partway through Herr Weyland's evening of hate, there was a sudden whispering and excitement in the hall. "Einstein, Einstein," the words moved through the auditorium and people turned to see where everybody else was looking. The great man had taken a seat in one of the boxes and showed himself confident as a king-bull walrus. Even the people who had come to scorn the Jew were eager to see the famous scientist in the flesh. Einstein had instinctively grasped the star's greatest weapon—the people's desire to look and report "I saw somebody famous." Weyland came in second at his own show that night.

Most of Berlin's scientific community had no respect or instinct for publicity and artificial fame. They voiced dismay that one of their own had dignified the musical-hall scandal with his showy presence. Einstein then expressed a thousand pardons to his colleagues, protesting that he had innocently thought it best to beard the lion in its den. Yet Einstein did not retreat into the academic monastery where his colleagues preferred to see him. Like Martin Luther, who shocked many clerics by writing in common German, Einstein then alarmed his friends again by defending his science in a newspaper, the *Berliner Tageblatt*. Science is not a democratic pursuit that progresses through majority opinion, and practical scientists could see no justification for arguing theoretical points in the popular press. Even more distressing to them was Einstein's razor-swipe at the throat of fellow scientists.

The most distinguished physicist in the anti-relativity movement was Philipp Lenard. His presence in the anti-Einstein party was even more tragic than Stark's. Lenard should have been Einstein's anchor in the experimentalist's party because light quanta rested on Lenard's experiments. The two men had once been friendly correspondents, but the war, and especially defeat, had scarred Lenard's brain. In Einstein's newspaper reply to his critics, he hacked at his opponent's public reputation, saying that Lenard "has so far achieved nothing in theoretical physics, and his objections to the general theory of relativity are of such superficiality that until now I had thought it unnecessary to answer them in detail."

German physicists, bearded faces and all, gasped in dismay. Even Max Born's wife sent Einstein a letter regretting that he had been "goaded into that rather unfortunate reply in the newspaper," and a bit later Max himself had the effrontery to write Einstein, "In these matters you are a little child. We all love you, and you must obey judicious people (not your wife)."

Einstein had done more than set the soiled linen out for common display; he had slapped many faces. Most physicists—most scientists, even most Nobel Prize winners—contribute little or nothing to theory. They gather facts and hit upon techniques, but they do not alter the axioms that govern their work. Physicists like Einstein, Bohr, and Planck were unusual figures for having changed nature's ground rules. Many proud scientists could ask themselves: if Einstein says that about Lenard, what might he say about me?

Despite their fatherly advice, Einstein defied his colleagues by making no public retraction. He even arranged for an open debate with Lenard at the annual meeting of the Society of German Scientists and Physicians. Einstein was going for a kill. He was right, of course. Lenard's understanding of relativity theory was superficial and trivial. How much more fruitful a discussion about light quanta would have been. Einstein could have used an experimental partner just then. Lenard was among the few physicists whom Einstein had admired in his early days and Einstein's first wife had studied under Lenard in Heidelberg. At about the time Mileva was there, a visiting Japanese physicist reported that Lenard's lab was "perhaps the most active in Germany." In 1901 Einstein wrote to Mileva, "I have just read a wonderful paper by Lenard on the generation of cathode rays by ultraviolet light. Under this beautiful piece I am filled with such happiness and joy that I absolutely must share some of it with you." The "generation of cathode rays by ultraviolet light" was the old name for the photoelectric effect. Einstein's letter refers to Lenard's original publication proposing that electrons—what Lenard called cathode rays—are knocked from atoms by the light that strikes them. Almost two decades later, Einstein still could have used Lenard's experimental vigor and imagination. He desperately needed to find some decisive experiment that would prove that quanta, not waves, caused the photoelectric effect.

Instead, Einstein scheduled a debate with Lenard that took place on September 23, 1920, in Bad Nauheim, a resort near Frankfurt in southwestern Germany. Before the Great War, Bad Nauheim had boasted an international clientele; aristocrats came from every European land and a few Americans were added for seasoning. The baths there were considered beneficial to people with weak hearts. A "heart condition" sounds alarming and dangerous, but vague heart conditions are as standard a diagnosis in Germany as a liver crisis is for the French, or a 24-hour bug is for Americans. Thus, despite its reputation as a spa for people with weak hearts, the visitors—who frequently stayed for weeks or even months—were not so much the old and infirm as they were the rich and leisurely.

It was not Einstein's kind of setting, not even in September 1920 when rich, leisurely foreigners were no more likely to visit Germany than they were to vacation across the Styx. Einstein stayed in Frankfurt with his friends the Borns for the meeting. Max Born had left Berlin to teach at the university in Frankfurt, so Einstein was spared the luxuries of the Nauheim hotels and their dining rooms where rolls were served in nickel-silver baskets. Each morning Einstein and Born rode the train 20 miles from Frankfurt to Nauheim.

The sessions revealed that besides the anti-Einstein movement, a more generalized anti-Berlin movement had gained strength in physics. Germany had many great cities besides its capital, and the provinces, which had been independent realms not so long before, were in a mood to assert themselves again. Leading the anti-Berlin struggle was Wilhelm Wien, winner of the 1911 Nobel Prize for work leading up to the quantum's discovery. Along with Max Planck, he was considered the grand old man of German physics. But Planck was disgusted with Wien and told a colleague that "Wien was motivated not by considerations of substance or the effect on science but by his anti-Semitic, right-wing attitude."

On the morning of the Einstein-Lenard debate, the police assembled in front of the Kursaal, the spa's main building, to make sure that only the meeting's registered participants entered. The armed guards no doubt disappointed the idly curious and any would-be troublemakers, but it was surely a blessing for them. It is difficult without grinning to picture nonscientists sitting through the Einstein session. Imagine some ill-educated nationalist who has come to boo the Jew, or a progressive come to cheer the democrat, having to sit still while Hermann Weyl, somebody they had never heard of, described his attempt—a completely mathematical attempt, reported with neither pictures nor flashy equipment—to unify Einstein's gravity with electricity. Presumably the tedium would have grown so intense that the interlopers would have lapsed into comas.

"The world is a curious madhouse," Einstein wrote his old schoolmate Marcel Grossman just before setting off to Bad Nauheim, "at present every coachman and every waiter argues about whether or not the relativity theory is correct. A person's conviction on this point depends on the political party he belongs to." When the coachman and the waiter disputed, they did not discuss the four- and even five-dimensional equations that theorists discussed on the floor of the Kursaal at Bad Nauheim.

At noon the main event, Einstein versus Lenard, began, with Max Planck in the referee's corner. Lenard spoke first. Einstein took notes, inevitably with a pencil borrowed at the last minute. Lenard had a lean and angular face with an old-fashioned, thin beard. Photos sometimes show a wild look in his eye, but that might have been a transitory phenomenon. One of the things that attracted the world-fame machine to Einstein was his well-proportioned face that made for pleasing, interesting photographs. Lenard, like most of us, was not so blessed.

Lenard spoke calmly, but he did not shy from including racial cracks in his presentation. His main semiscientific argument was that Einstein's theory defied common sense. Einstein pooh-poohed the objection. Common sense changed with the times. Newton's ideas, which Lenard defended as pure clarity, had once seemed difficult because they challenged Aristotle's common-sense positions. And Aristotle, too, had once seemed arcane and remote. Saint Augustine recalled Aristotle as the most difficult person he had studied when he was a schoolboy of the Roman Empire. Yet it was not so easy to refute Lenard because the importance of common sense comes down to personal taste. Whose authority do your instincts favor? Einstein and Lenard, as scientists, were supposedly agreed that observation trumped both logic and common sense. However, Lenard insisted that observations did not yet support Einstein, dismissing with banalities and a few anti-Jewish remarks, the successful observations of the eclipse that had hurled Einstein onto the world stage.

In their light-quanta dispute, stalemate had pestered Einstein and

Lenard for years. Lenard had said it was common sense that light waves were at work in the photoelectric effect. Light waves worked on electrons like ocean waves on beach balls. A wave crashes against the shore and sends balls rolling inland. The harder the wave hits the ball, the harder the ball goes rolling. If the wave frequency increases, the balls will bang around more. Lenard had done extensive work studying light's effect on electrons and demonstrated that this beach ball analogy did not work. The electrons do not gain energy when the light grows more intense, nor do they bang around more when the frequency increases. Instead, light gets things backward. Increasing the light frequency—not its intensity—increases the electron's energy. Increasing the light intensity, not the frequency, increases the number of electrons knocked free.

Aha, said Einstein, the ocean wave analogy does not hold. Light is not a wave. Lenard and every other established physicist looked toward the stick's other end and said there must be something very peculiar about electrons for them to be reacting as they do. Perhaps the long frustration over this ambiguity finally caught up with Einstein. At Bad Nauheim he suddenly lost his temper and snapped at Lenard. Despite the meeting's police barricade, some demonstrators had found seats and chanted slogans as Einstein spoke. Einstein kept talking. Planck turned white as marble. Finally Planck made a feeble joke, "Since the relativity theory unfortunately has not yet made it possible to extend the absolute time interval available for the meeting, our session must be adjourned." And the debate ended.

Einstein's triumph was not at once clear. Only one month spanned the time between his attendance at Weyland's show, his newspaper article attacking Lenard, and his appearance at Bad Nauheim. Einstein's willingness to go public robbed the anti-relativity movement of its scientific pretensions before it was well away. Weyland soon fell back into obscurity, not to be heard from again until 30 years later when he volunteered as an informer against Einstein for America's Federal Bureau of Investigation. Einstein had held his ground. His enemies could not repudiate his science, although they continued hating him as a Jew, a traitor, and a dreamer.

8

Such a Devil of a Fellow

he anti-relativists were partly parasites grabbing some free fame like barnacles stealing a free ocean voyage, but they were also fuel that kept the fame machine turning. Industrialized fame is wind powered; its blades turn only when something blows. The overthrow of the Newtonian universe had swept the hats off many unsuspecting heads, but eventually the word spread and that breeze began to fade. Then the anti-relativists appeared and the fame engines had something further to turn them, but Einstein's effectiveness at facing down his challengers made anti-relativism look like a losing cause. Weyland's secret financiers appear to have halted his funding, and new backers did not step forward. Once again the fame mills seemed to be winding down for Einstein. Then in 1921 he began to travel widely and the fame machinery resumed its steady whir.

In March 1922, a year into their goings and comings, Einstein and Elsa hosted a dinner party in their Berlin apartment. The guests included some of the city's most fashionable and wealthiest people. Members of the Warburg banking family attended. Among the fashionable guests was Count Harry Kessler, a diplomat who knew everybody, went everywhere, and wrote it all down in his diary. He had spent the afternoon with the foreign minister, Walther Rathenau, trying to find a way for Germany to again figure in international affairs.

Then, in the evening, he was at Einstein's where the topic of the hour was the amazing ease with which Einstein had penetrated every foreign heart. Einstein described for his guests his triumphal receptions in America and Britain, but he admitted that he could not grasp why people were so enthusiastic.

Einstein, of course, understood what was new in his theory of relativity. Previously, people had thought that matter, space, and time were separate, fundamental things. He showed that they are like the corners of a triangle. Each has meaning only in the presence of the other two. But why, Einstein wondered, did people become so excited about that? What change did relativity make in the way people looked at their own lives? In Britain the Archbishop of Canterbury told Einstein that he had heard relativity "ought to make a great difference to our morals." "Do not believe a word of it," Einstein reassured him. "It makes no difference. It is purely abstract science."

Elsa told Kessler that her husband felt like a conman who arrived to great applause and then failed to deliver on his promise. Pre-science had been filled with figures who, if not full confidence men, enjoyed and gained authority from their impenetrability. Paracelsus, the Swiss-born alchemist, is remembered both for his insistence on experiment and his obscure language. He coined pseudoscientific words like "alkahest," choosing them for their Arabic sounds to enhance his authority and learned appearance. Two generations later Giordano Bruno managed to get himself excommunicated by the Lutherans and burned as a heretic by the Catholics. One thing seems clear: nobody understood exactly what Bruno taught in his treatises on magic and occult astronomy. Yet his obscurity provoked great emotion and excitement among the baffled. Einstein, who spent his whole life struggling to understand, was dismayed to find himself cheered for ideas that the hurray-sayers could not grasp.

Kessler confessed to sensing the significance of Einstein's thought more than to understanding it. It was exactly the response sixteenth century readers had had to Paracelsus and to Bruno. Einstein was a champion for all those who hoped the world made sense but could not make sense of it themselves. By being smarter than them, Einstein had seen the way the universe does indeed make sense. With his successful stride toward understanding he had assured the halt and the lame that the fault is in themselves and not in the stars.

Berlin's distinguished visitors commonly included Einstein on their list of must-sees. The Indian poet Rabindranath Tagore came to Berlin in 1921 and sought out Einstein. Although such an East-West, art-science, religion-secularism meeting strikes some tastes as absurd, both men saw themselves as separate points in a shared, human civilization and they got along. Tagore's friends called him Rabi. Einstein jokingly added another "b" and called him Rabbi.

Whenever he traveled outside Berlin, Einstein was cheered as the good man who knew what was what. He went first to Prague, a city where he had taught before the war, and then to Vienna. Next he sailed to the United States. On the way home from America he stopped in Britain to become the toast of London. Everywhere he went, the local machinery rolled out more applause and more fame. There were banquets, lectures, and crackpots. People snapped photographs and shouted questions. Elsa came along and learned how to play the role of simple goodheart married to the great man. In New York she laughed off American reporters who wanted to know if she understood her husband's theory. In London, when the Archbishop of Canterbury's wife said that a friend had been explaining the mystical aspects of Einstein's theory, Elsa burst into laughter, "Mystical! Mystical!" She was having a roaring good time. "My husband mystical!" and she kept laughing.

Democratic Germany was delighted to find that it had a popular goodwill ambassador, and its diplomatic pouches carried home the good news of Einstein's friendly receptions. Even back in Germany Einstein was turned into a make-believe traveler and fêted at a grand party, just as though he were a foreign dignitary honoring Berlin with a visit. The republic's cabinet ministers, dressed in their bourgeois best, turned out with their wives. President Ebert also came to meet Einstein, giving each man the chance to see how the other had changed since the afternoon when Ebert had given Einstein a chit to procure the freedom of the university's rector and faculty.

Einstein seemed to love travel. Between his journey to Vienna and the United States, he made a brief, diplomatic trip to Amsterdam and enjoyed his first ride in a sleeping car. The diarist Count Kessler traveled with him and was astonished to see a middle-aged man become as wide-eyed and curious about everything as a child on his first train. Sea voyages fascinated him still more, and passengers noticed his curiosity about the machinery used to keep tables steady on a rolling sea.

Wherever he went, lunatics and well-meaning naïfs were eager to talk to him. The cranks whom Einstein especially regretted were the mad warriors who wanted to turn relativity into a bomb. After Einstein showed that, like ice and steam, matter and energy are different forms of the same thing, people began to wonder what would happen if, on cue, matter melted into energy. In 1914, H.G. Wells published a story about a bomb that converted atomic matter into explosive energy. Amateur scientists often wrote to Einstein about such ideas, seeming not to realize that the pacifist Einstein was revolted by their ambition. In Prague a would-be bomb maker managed to corner him briefly. As Einstein made his escape he told the young man, "You haven't lost anything if I don't discuss your work with you in detail. Its foolishness is evident at first glance."

All this travel and hectoring was a great distraction from his scientific work. Light, quanta, and the link between gravity and electromagnetism were always bubbling in the corners of his mind, but it was hard to practice science while living in the public view. He had been looking forward to the renewed Solvay conference and the chance to put science first. The original Solvay conference, 10 years earlier, had been one of the outstanding scientific gatherings of the twentieth century. Lorentz had presided, as he did at succeeding Solvay conferences. Einstein had presented a paper. Marie Curie had been there. Planck had come. The conference had been organized to alert the scientific world to the way Planck's quantum was revolutionizing everything. The conference made its point. After it was over, Henri Poincaré went into a creative frenzy and proved mathematically that the quantum hypothesis was the only possible explanation for the observed facts.

That first Solvay conference had been inspired by German physicists and chemists who wanted to publicize a German idea—the quantum of energy. They had persuaded a Belgian millionaire, Ernst Solvay, to give his money and name to create the conference. A second had been held in 1914, just as war was breaking out. The 1921 conference marked their resumption, but Einstein was now the only German invited. His colleagues in Berlin were angry and insisted that Einstein should not attend. All of Einstein's internationalist leanings supported going. Also, it was absurd that he could travel around the world in his role as famous man, but not as a working scientist. Yet, he knew German science was still great science and that serious international science could not do without it. So, as a gesture to German science, he did not go to Solvay 1921, politely explaining that he would be touring America instead.

As it turned out, the third Solvay conference was not as important as its predecessors. Without the Germans, how could it be? Bohr did not attend either. Quantum physics was stuck. The successes that followed the first Solvay had faded during the war. It was clear that physics needed a new idea, but physicists were reluctant to embrace anything really radical. Einstein's light quantum, with its inherent particle and wave properties, was the most successful radical idea on hand, and yet few physicists, apart from Einstein himself, dared be that revolutionary. Accepting light quanta would mean reconsidering nineteenth century physics' greatest achievement, establishing that light was an electromagnetic wave. Newton had taught that light was a substance composed of particles, but early in the nineteenth century a physician named Thomas Young, one of the Rosetta Stone's translators, had performed a series of experiments that could only be explained if light were a wave. The French mathematician (and yet another Egyptologist), Baron Fourier, developed a technique for analyzing wave motions. Physicists could suddenly make very precise calculations to test the wave theory of light. James Clerk Maxwell had finally clinched the argument when, by simply assuming that light is a wave, he introduced a wonderfully accurate series of equations that unified light, electricity, and magnetism. Like nervous dukes on the eve of reckoning, physicists could see that something drastic had to be done if they were to stave off revelation, but they were unwilling to defy past achievements. So Solvay 1921 came and went without clarifying quanta, atoms, electrons, or the photoelectric effect.

There was a moment in that summer of 1921 when something new might have been tried in quantum physics. In Copenhagen, Bohr's chief assistant, Hendrik Kramers, thought he had found a way to test Einstein's light quanta idea and resolve the dispute for good and all. He noticed that if a quantum loses some energy it must change to a lower frequency. The math behind Kramers' idea was extremely simple. The energy of a quantum equals Planck's constant (h) times its frequency (v). Since h never changes, any change in energy must reflect a change in v. Thus, if a quantum of blue light loses energy, the light will move down the spectrum to some other color, perhaps red. Red light, when it loses energy, might even become invisible as its frequency drops down to the infrared band, which the eye cannot detect. But how could a quantum lose energy? One way, if you take Einstein's quantum seriously, would be for it to collide with something and transfer some energy just as one billiard ball loses energy when it collides with another.

Kramers was intensely excited to realize that he could establish or abolish Einstein's idea, but Bohr pointed out that the idea assumed that energy conservation holds on the quantum level. Could Kramers really be sure that energy changes for quanta would really work the way billiard balls worked? Quantum energy conservation had not yet been proved and, personally, Bohr did not believe in it. Kramers disagreed. His idea was beautiful and should be tested. Bohr argued with the tenacity of an interrogator for the secret police. He reasoned, he threatened, he cajoled and insisted. Kramers stood his ground, then wavered, then grew sick. Bohr nursed him and kept up his interrogation. Finally Kramers saw reason and abandoned his hope for a decisive experiment.

It sometimes happens that interrogators do more than wheedle a confession from somebody. They break them and their humanity just as a determined rider can break a wild horse to the saddle. Something like this seems to have happened to Kramers, because after this duel with Bohr he changed. Visitors to Copenhagen after that summer were often surprised by what a tame and compliant assistant Bohr had in his Kramers.

The fame machine never pressed Einstein on this light–quanta business. Everywhere he went he was asked to explain relativity, but never to discuss his other ideas. Industrialized fame was clouding Einstein's successes, reducing his whole achievement to that one incomprehensible word—relativity. Often, reporters said that he had revolutionized science, and usually he denied the claim, insisting that relativity was merely an extension of established physical ideas. Apparently no reporter thought to ask, "Well, have any of your other ideas been revolutionary?" Perhaps they did not know he had had other ideas.

At least not other scientific ideas. Einstein did make one big unscientific pronouncement that caught the public's ear. He deepened German nationalist contempt and he horrified intellectual German Jews by announcing that he was a Zionist. Einstein's trip to America, although it included many scientific lectures, was billed as a fund raiser for the proposed Hebrew University in Jerusalem. He told a Jewish colleague in Berlin, "Naturally I am needed [in America] not for my abilities, but solely for my name, from whose publicity value a substantial effect is expected among the rich tribal companions in Dollaria." Inevitably, Einstein's proposed travel to wartime enemy countries drew yelps from German nationalists. Many assimilationist Jews were troubled as well. Adolf Hitler, at that time an unknown, would-be rabble-rouser, wrote an editorial in his party newspaper at the year's end warning that "science, once our greatest pride, is today being taught by Hebrews . . . [as a means of] triggering the inner collapse of our nation." At the time it seemed only more noise in the fame machine's persistent hum.

If Einstein had won the Nobel Prize that year he would have spent December in Stockholm, but no prize in physics was announced for 1921 and Einstein passed the year's end in Berlin doing science. Despite fame's distractions, he continued to think about physics. He was not required to teach, but when he was in Berlin he liked to participate in a weekly colloquium at the university. The students appreciated these sessions because of Einstein's immediate ability to find the central issue of any idea. Perhaps his Swiss heritage was coming to the fore. Like William Tell, his arrows seemed always to strike their targets. Tourists sometimes came to these sessions as well and listened as Einstein spoke in a language they did not know, about questions they could not imagine.

The physics students were glad of these sessions, but others were not always so taken with them. The revolutionary students who had seized the university immediately after the war's end had been a kind of mirage that soon disappeared from the postwar scene. Most of Germany's youth was not socialist, not bohemian, not democratic, not even political. They were students with the general concerns of their mates around the world—classes and sex. The minority of students who considered larger issues leaned toward the political right, even the ultraright. As early as 1920 a national student convention meeting in Göttingen voted to limit their membership to "Aryans," that is, no Jews allowed. This was more than a dozen years before the Nazis came to power. Generally, their teachers were not likely to challenge them. History, literature, and languages professors especially were heavily antidemocratic, anti-communist, anti-Jew.

Einstein did his own work off campus, mostly in the study room above his apartment. Then he would report the results at the next meeting of the Prussian Academy. Following this procedure late in 1921, Einstein finally hit on an *experimentum crucis*, to use the old Newtonian term for the decisive experiment that settles the argument between two ideas. Einstein had finally seen a way to distinguish between wave effects (which travel spherically) and quanta (which move in one direction). It was a way quite unlike the one dreamt up by Kramers. Einstein had designed a complicated device that focused beams of atoms and then refocused them to produce a second, parallel beam. Einstein calculated that if light travels in quanta there will be nothing extraordinary about the second beam, but if it travels in waves, the second beam's wave length should change and the beam should fan out a bit.

Decisive experiments add the gambler's reckless thrill to theoretical science. Years of thought ride behind one spin of the wheel. Einstein had worried that he might never find an experiment to test his theory of gravity against Newton's; finally he realized that a total eclipse of the sun could be put to use. He had struggled even harder to find some test subtle enough and decisive enough to settle the light quanta issue. Ehrenfest sent Einstein a note, "If your light experiment turns out anticlassically [that is, by disproving light waves] . . . then, you will have become really *uncanny* to me." He added that proving light travels as quanta rather than waves would be "something completely colossal."

Two experimentalists quickly built a working apparatus to create and measure Einstein's beams. The wheel spun and up came the measurements. The wavelength did not change; the beam did not spread. Once again Einstein's numbers were on the money. He wrote to Max Born, "This has been my most impressive scientific experience in years." At last he had a chit proclaiming his light quanta's authority.

The fame machine made not a peep, although perhaps that was just as well. Most classical theorists were out of their depth when it came to judging Einstein's reasoning, but Einstein's closest friends and colleagues understood what he was saying and they disagreed with him. Max von Laue energetically challenged Einstein in the Prussian Academy, and two days after his enthusiastic note, Ehrenfest sent Einstein a letter saying that he had seen a flaw in the calculations. Einstein, he said, had missed the point that in this experiment the wave behavior would match that of light quanta. Wave or quanta, the test would give the same results. Ehrenfest ended, "Of course you are such a devil of a fellow that naturally you will finally turn out to be right in the end," and he begged, "Don't be annoyed with me if I am wrong; and don't be annoyed with me if I am right."

Einstein was annoyed. He redid his calculations, this time with more rigor and told Ehrenfest that the recalculation still supported him. A week later, however, he wrote Ehrenfest, "You were absolutely right." Further analysis had shown that in this experiment waves and quanta would behave in exactly the same way, so his "decisive" test could decide nothing. The chit was worthless.

Again the fame machine was quiet, at least about this matter. It had more amusing elements of the Einstein story to hum about. While Einstein was working on his experiment, the press erupted with a supposed quotation in which Einstein said that American men "are the toy dogs of the women, who spend money in a most unmeasurable, illimitable way and are themselves in a fog of extravagance." Einstein protested that he had been misquoted, but masses of articles, editorials, and letters to the editor held the public eye. "I suppose it's a good thing I have so much to distract me," Einstein wrote Ehrenfest, "else the quantum problem would have long got me into a lunatic asylum." But distraction was taking its toll. It is hard to believe than an Einstein left fully to himself, working in his study with Elsa guarding the door, would have announced his misguided "decisive" experiment.

Einstein then went off to France where he was booed as a German and cheered as a genius. It looked as though 1922 was to be like 1921, a year of mad publicity, but on June 24 the foreign minister Walther Rathenau, a German patriot who had been instrumental in finding the money that kept the Kaiser's army in the field, Einstein's friend, and also a Jew, was assassinated by German nationalists. Einstein was rumored to also be on the list of Jews to be killed immediately and he disappeared from public view.

9

Intuition and Inspiration

The murder of Walther Rathenau had a look that was to become increasingly common in the twentieth century. The foreign minister was riding in an open car when another automobile pulled alongside and the occupants began blasting. The two cars swerved, one from panic, one from hatred. The shooters preached the text that underlies all thug politics: *Shaddap. This means you. We can do without your ideas.* Rathenau had been warned frequently that there were plans to assassinate him, but he refused to take them seriously. Shortly after that killing an attempt was made against another prominent Jew, Maximilian Harden, a journalist, essayist, publisher, and radical socialist. Although Harden survived his attack he retired from public life and left for Switzerland. Force was replacing rumor as the source of authority in Berlin.

The story of science, art, and other fruits of imagination is more burdened with efforts to silence creators than we commonly recognize. Although silencing, say, a physicist because he is a Jew, and a pacifist, and a democrat makes the physics part seem accidental, it is the general open-mindedness necessary to the successful creation of ideas that makes thugs so angry. A British mob in Birmingham made clear their willingness to do without further thought from the chemist, oxygen's discoverer, and religious radical, Joseph Priestly. They hated him for his broadmindedness, not for his oxygen, but the broad-mindedness led to oxygen. Thugs stormed his home and destroyed his library, his laboratory, and the rest of his place. *Shut up, shut up, be silent* was the message that sent Priestly fleeing into the night when thugs thought he was too sympathetic to the French Revolution.

Across the channel, in Paris, the man who explained the importance of Priestly's oxygen was also spotted by thugs. They saw the chemist as the revolution's enemy and architect of the wall that forced all goods to enter Paris through tax collection points. They hated Lavoisier for his cleverness, not his theory of fire, but of course the cleverness was foundation to his theory. *No more from that brain* was their thought as they chased him down the city streets.

Einstein ducked down an alley, too, and stayed quietly out of view in his apartment at 5 Haberlandstrasse. He had never lacked for scientific or political courage, but he was not foolhardy. With brutes roaming the boulevards, he lay low. Yet guests did come over to play music with him. Professional musicians enjoyed playing with the famous host. His apartment featured a music room where he and guests would play while Elsa listened through open double doors. Einstein loved his violin, playing the notes straight, with no vibrato. "Skreek-skreek-skreek," was the way the pianist Rudolph Serkin recalled the sound, and Einstein loved every minute of it. It was as distracting and seductive as fame, a pleasurable way to lay by his quest for sense. If Einstein ever idolized anything—in idolatry's formal sense of worshiping a human creation—it was music.

In the noisy, boisterous world beyond Einstein's apartment, a whole civilization was crossing a Rubicon. Anything-goes open-mindedness was going to war with ferocious hatred of the different. In poetry, 1922 was the year T.S. Eliot published *The Wasteland* and Rilke, "in a creative fit," as Peter Gay described it, scribbled down his *Sonnets to Orpheus* and the *Duino Elegies*. In prose, Joyce's completed *Ulysses* was printed and another volume of Proust's great work rolled off the presses. Musically, it was the year Louis Armstrong quit a riverboat band to settle in Chicago and begin transforming jazz into an art form. Meanwhile, hatred for all such mental liberties was organizing. On the day Rathenau was murdered, Canadian readers awoke to find two stories about Italian fascism in the morning's *Toronto Star*. The reporter was Ernest Hemingway. The first story related an interview with Benito Mussolini, who told the newspaper representative, "We are a political party organized as a military force." Hemingway's second story described the members of this military force as "black-shirted, knife-carrying, club-swinging, quick-stepping, nineteen-year-old potshot patriots." As for the suggestion that anything more than banality underlay fascism, in a later report Hemingway wrote, "Mussolini is the biggest bluff in Europe. If Mussolini would have me taken out and shot tomorrow morning, I would still regard him as a bluff. . . . Study his genius for clothing small ideas in big words."

While Einstein stayed hidden, the struggle between big ideas and big words took shape. Einstein's enemy in physics and politics, Philipp Lenard, was unmoved by Rathenau's murder. The German Republic proclaimed a day of mourning for its slain minister, but Lenard said the Jew had been "justly" killed and he insisted on giving his regular physics lecture.

Hindsight makes it seem odd, but Germany in 1921 was seen by some as a haven from thug politics. For years Berlin had been absorbing refugees fleeing the Bolshevik catastrophe. Jewish peasants had recreated a shtetl on Berlin's outskirts while Russian nobility settled in Berlin proper. Vladimir Nabokov was among the refugees who thought Germany looked better than home, and who knew personally that murder had become a popular tool for shaping the future. Nabokov's father had been slain that spring at a public meeting of Russian émigrés. The question before the floor had been what to do about communism's successful entrenchment back home. During the evening two monarchists tried to shoot an ardent republican. They were poor marksmen and killed Nabokov senior instead. The battle between openness and close-mindedness had become a war.

Open-minded Germany still reigned on the day of the Rathenau murder and the country's middle class was mostly horrified by the assassination. The people's indignation made the nationalist movements fall back on the defensive, and naturally some on Germany's political left called for silencing the right. The chancellor told the Reichstag, "The enemy is on the right—here are those who drip poison into the wounds of the German people!" and he began introducing legislation to "protect the republic" by silencing extremists. The world had become a remarkable garden in which some people were creating unknown flowers while others were determined to rip out all new blooms as though they were weeds. Blows from all sides followed. In Heidelberg, on Rathenau's day of mourning, parading workers saw the lights in Lenard's classroom and, knowing nothing of his thoughts about the photoelectric effect, burst into his class, grabbed him by his clothes and frog-marched him out the building. They were prevented from tossing him off a bridge only by the intervention of more humane mourners.

How, faced with such murder, was open-minded expression supposed to triumph over narrow-mindedness organized as a military force? All these generations later, of course, the question is hardly suspense-filled. We know that the close-minded militarists were finally defeated by the creative societies, which were slow to rouse, but unstoppable once angered. We know, too, that fascist ignorance paid a terrible price for embracing silence. Many of the dreamers in this story would eventually move abroad, where they contributed their talents to the other side. Einstein famously settled in Princeton, New Jersey, where for a time his neighbor was the refugee-novelist Thomas Mann. Max Born would end up teaching in Edinburgh. Ilse Rosenthal-Schneider, the philosophy student who chatted on the tram with Einstein, spent the last half of her life in Australia. Erwin Schrödinger, who figures prominently later in this story, went to Dublin, while the Dutchman of great merit, Peter Debye, skipped to Cornell in New York State. The musicians who played in Einstein's music room typically ended up in New York City or London. Even Bohr would make a daring wartime escape from occupied Denmark to America. Meanwhile, the fate of the university at Göttingen showed what happens when banality takes charge. The university's mathematics school had looked like a kind of perpetual-motion machine that spewed out ideas in an endless show of fireworks, but Göttingen ran out of fuel as soon as the Nazis put an end to the imaginative life.

The long view, however, misses the fears of daily life. In the summer of 1922 thuggery was getting away with murder, literally. Imaginative people were being killed and silenced. German, Austrian, and some neutral scientists were banned from international conferences. Rathenau's murder shook Einstein to the depth of his soul. It foreshadowed a long assault on Jews and humanism, and Einstein knew it. At the same time, Bohr was troubled by the larger world's suppression of German science. Isolated Germany was developing its own distinct theory of the atom and quanta. The first Solvay conference had moved quanta to the heart of European physics, but time had eroded that universality. Poincaré died in 1912, before his newfound enthusiasm for the quantum revolution could take deep root in France. Britain's leading quantum scientist was a brilliant physicist named Henry Mosely, but British interest in quanta faded after he died violently on the Turkish front. Only in Germany had the pace and breadth of work with quanta accelerated.

So Bohr made a one-man show of openness toward Germany. When Rathenau was assassinated, Bohr had just completed a triumph in Göttingen. Over a two-week period he gave seven lectures at the "Eldorado of erudition" as Einstein jokingly designated the university. It was a great event for the school, the students, and the scientists who remembered those two weeks as the Bohr Jamboree, or Bohr Speakathon, or however you choose to translate *Bohr Festspiel*. The most promising young physicists in Germany made their way to Göttingen to hear Bohr, partly because he was a great physicist and partly because he was a great physicist who had come to Germany when many *Auslanders* would not even shake hands with men like Max Born, Arnold Sommerfeld, and Max Planck. Bohr knew that his broadmindedness about ideas stamped "Made in Germany" kept him in the game. An alliance was developing that linked the scholars of Copenhagen, Munich, and Göttingen in a shared approach that took the Bohr atom for granted and saw no use for Einstein's hypothetical light quanta.

Einstein returned the compliment by being skeptical about the intricate details of Bohr's atom. True, its experimental success was impressive and surely must mean something, but there were still many contradictions and unsolved puzzles. Eighteen hundred years earlier the astronomer Claudius Ptolemy had said, "Theoretical [physics] should rather be called guesswork than knowledge . . . because of the unstable and unclear nature of matter." All these centuries later the nature of matter was still unclear. Bohr had made some brilliant guesses—intuitions, his disciples called them—but was he right?

Ptolemy himself had described the heavens, and his method allowed for wonderfully accurate calculations of eclipses and astrological relationships. His theories were developed when the physical nature of the night sky was almost a total mystery. It consisted of a show in which lights moved and varied in brightness. Everything known or guessed about the structure of the cosmos came from carefully observing and measuring the motions and changes in the brightness of those lights.

Didn't they know that the stars and planets were places, separate objects with their own landscapes or features? No, not really. As late as Galileo's time the Holy Roman Emperor was asking the court astronomer if he thought the moon might be a mirror that reflected part of the earth's surface. The ancients knew only those moving lights and their changes in brightness. Beyond that it was all "Twinkle, twinkle little star, How I wonder what you are."

How much of Bohr's atom was as fictitious as the crystalline spheres that were once said to carry the planets around the earth? The only thing anybody really knew about atomic behavior concerned the frequency and intensity of the light that atoms absorb and emit. Bohr talked about an atom's orbiting electrons and the distortions of these orbits caused by electric and magnetic fields, but these theoretical motions were based on observed changes in the frequency and intensity of atomic light. Nobody had seen an electron moving around an atom.

Yet Bohr was building a great following in Germany. Physicists to the political left of Lenard and Stark packed his lectures and heard the Danish visitor explain that chemistry's periodic table of the elements made sense. Fifty years earlier, Dmitry Mendeleyev had produced a chart that organized the elements into groups with shared chemical properties. Mendeleyev's table, however, was much like an inexplicable equation or system of categories. It had its uses, but only after Bohr's lectures did it become clear that quantum physics suggested a meaning that lay behind the periodic table. Bohr's idea was that electrons can share orbits so that an atomic "solar system" has several "planets" following the same route—as though the earth had a twin sailing the same course around the sun. Bohr said that the number of electrons in the outermost orbit determines an atom's chemical properties and he showed that chemicals listed in the periodic table as being similar shared the same number of outer-orbit electrons. It was a wonderful example of finding physical meaning behind factual order.

It is entertaining to speculate on how many people in the audience actually understood Bohr's lectures. Despite his brilliance, he was remembered as a "divinely bad" speaker. His voice was weak and carried only to the front rows. Even worse, he was rethinking his way through his lectures as he spoke, so nothing came trippingly or directly. Audiences were commonly more bewildered than enlightened by his words.

The smartest quantum people, however, knew what he was talking about. Werner Heisenberg was a bright student in Munich who had come to Göttingen just to hear Bohr. He reported, "Each one of [Bohr's] carefully formulated sentences revealed a long chain of underlying thoughts, of philosophical reflections, hinted at but never fully expressed. I found this approach highly exciting; what he said seemed both new and not quite new at the same time. . . . We could clearly sense that he had reached his results not so much by calculations and demonstrations as by intuition and inspiration, and that he found it difficult to justify his findings before Göttingen's famous school of mathematics."

Bohr's distaste for math was his most distinctive trait. It is common in populations as a whole but unusual among theoretical physicists. He did not like the way mathematics encouraged people to believe abstractions were real. Mathematical formulas tend almost inevitably to propose abstractions underlying appearances, while natural languages force ideas to stay closer to the surface, and Bohr wanted to understand surfaces. Einstein was the opposite, treating his abstractions with utmost seriousness and even saying that hv, which combines an uncountable number (h) with an abstract property (v), is something that actually bangs about and causes visible results.

Oddly, Bohr's persuasiveness might have rested in part on his mathematical limitations. Although he cited equations throughout his lectures, there was very little mathematical reasoning of the kind in, say, Einstein's original paper on light quanta. There, if you cannot follow the math you cannot follow the argument. Instead of undercutting Bohr, this mathematical thinness made him seem more profoundly connected to physical reality. Of course, his admiring listeners did have an instinct for mathematics. Thus, even though he found Bohr's lectures exciting, Heisenberg rose to challenge one of the calculations Bohr had reported, and Bohr had no good reply.

Bohr was frank about the many mysteries that atomic structure still posed. The fundamental challenge lay in tracing what happened when two or more electrons move through subatomic space. It was proving impossible to distinguish even the theoretical motions of one electron from another. New refinements seemed constantly required as experimentalists discovered new subtleties in light intensity and frequency. For Einstein, these problems meant that the basic theory of quantum radiation was, at best, incomplete. For Bohr, the fundamentals were sound; physics just needed more tools. Bohr was out there proselytizing, winning interest while Einstein stayed invisible, working in his study or skreek-skreek-skreeking on his violin.

10

Bold, Not to Say Reckless

Who are the three most significant people alive? A Japanese publisher asked Britain's already well-known philosopher
Bertrand Russell that question in 1921.

Einstein and Lenin, Russell answered. Perhaps a sudden fit of mod-
esty led him to omit a third name. The mills of the worldwide fame
machine, Japanese branch, began to turn and Einstein signed contracts
calling for a dozen lectures during a six-week tour of Japan. After the
Rathenau murder, Einstein canceled all his scheduled lectures and public appearances, except for the Japanese plans.

In September, while Einstein still hid from public view, he received word that he should not do anything that would prevent a trip
to Stockholm in December. It seemed that the Nobel Prize was coming. However, Einstein refused to postpone the journey to Japan, partly
because he had signed a contract, partly because prizes, even Nobels,
did not mean much to him, perhaps also because the prize money was
already promised to his first wife, Mileva. Einstein's action carried
some risk. The Nobel committee was easily offended and might not
vote an award to someone who did not appreciate its grand importance, but by autumn 1922 the shame of Einstein never having received a Nobel fell much more on the prize givers than it did on the
supposed honoree.

Einstein and Elsa made their way discreetly from Berlin to Switzerland, and then to the French port of Marseilles where a Japanese passenger steamer awaited them. It was an excellent time to be leaving Germany because the mark had begun to collapse into worthlessness. The life savings of millions were being consumed as though by fire.

It took six lazy weeks at sea to reach Japan, giving the passengers ample time to get used to having so renowned a traveling companion. Toward the journey's end, as the ship traveled from Hong Kong to Shanghai, Einstein received the news that the previous year's Nobel Prize in physics had at last gone to him. At the same time the Swedish Academy announced that the 1922 prize was awarded to Niels Bohr. There was some confusion as to just what Einstein won the prize for. The award statement seemed to indicate that his theory of relativity still held uncertain status, then it mentioned Einstein's general "services to the theory of physics, and especially for his law of the photoelectric effect."

Mondo bizarro indeed! By 1922, only envy-maddened physicists were still questioning Einstein's 1905 theory of relativity. Those who knew enough to have an opinion also accepted his view of gravity. Most physicists, however, continued to reject light quanta as an explanation for the photoelectric effect. Of course, the committee did not award the prize for the light quanta, only for the photoelectric "law." That law states that the energy of an electron sent flying by electromagnetic radiation equals $h\nu$ minus a fixed amount of work needed for the electron to escape its atom. In Einstein's theory, $h\nu$ is a light quantum. The committee seemed to be ignoring the equation's meaning and mentioning only that the equation succeeded.

It was not even Einstein who had established his equation's accuracy. That achievement went to an American, Robert A. Millikan, at the University of Chicago. He began his experimental study of the photoelectric effect in 1905, the same year Einstein published his light-quanta hypothesis. The distance between Chicago and Berlin was very great in those days, and Millikan was not immediately aware of Einstein's theory. If he had been, he would not have been impressed, because Einstein's law has only one variable, the electromagnetic frequency. Millikan expected to find that temperature is most important in determining photoelectric action. As an American, who no doubt had enjoyed his share of popcorn, it is easy to picture what Millikan had in mind. Roasting popcorn's bursting rate and ferocity increases along with the rising temperature. He expected to find that the thermometer went up as electron energy increased. His experiments, however, determined decisively that temperature has nothing to do with the case.

Millikan's first experimental efforts paid two rewards. They removed temperature from any further consideration of the issue and they taught Millikan powerful techniques for producing and measuring the photoelectric effect. He learned how to test the effect with different materials and different wavelengths under many different conditions. There seems no question that he passed Lenard in his experimental knowledge of photoelectricity and its effects. The puzzle was that none of Millikan's experiments challenged Einstein's law.

In 1912 Millikan visited Europe and its leading physicists, including Planck in Berlin and Lorentz in the Netherlands. Neither had any sympathy for the light-quantum hypothesis. Following his return to America, Millikan published an article on the nature of radiation. After surveying four other theories, he reported, "The facts . . . are obviously most completely interpreted in terms of . . . [Einstein's] theory, however radical it may be. Why not adopt it? *Simply because no one has thus far seen any way of reconciling such a theory with the facts of diffraction and interference so completely in harmony in every particular with the old theory of ether waves.*" [Millikan's italics]

Millikan went back to work, testing Einstein's equation and by 1915 had proved that it was right on the money. The frequency of the electromagnetic radiation is the only property that determines an electron's energy when it is knocked free. Millikan drew a graph showing Einstein's prediction and another showing what he found experimentally. The match was so precise that it might have seemed suspicious if people had not known that Millikan was hoping to disprove Einstein. Millikan's experiments destroyed all rival photoelectric theories but did not establish Einstein's idea. Millikan insisted that accepting Einstein would be a mistaken response to his work. "The semi-corpuscular theory by which Einstein arrived at his equation seems at present to be wholly untenable," he wrote, adding that Einstein's "bold, not to say reckless, [hypothesis] seems a violation of the very conception of an electromagnetic disturbance."

Einstein was completing his work on general relativity when Millikan finished that photoelectric paper. Only after he settled relativity did he turn again to light quanta. Defiant as ever, he wrote a classic paper that is best remembered these days as laying the foundation of laser technology. Meanwhile, Millikan was still perplexed. In 1917 he wrote that Einstein's photoelectric law "stands complete and apparently well tested, but without any visible means of support. These supports must obviously exist, and the most fascinating problem of modern physics is to find them. Experiment has outrun theory, or better, guided by erroneous theory it has discovered relationships which seem to be of the greatest interest and importance, but the reasons for them are as yet not at all understood."

Seeing such universal doubt among capable physicists, it seems natural to ask why Einstein remained so confident of his light quanta. It was not that he had gone chasing after a windmill. He had done the math and found that Maxwell's theory did not work when applied to the facts of the emission and absorption of radiation. Using Maxwell's theory, Einstein proposed an equation—we can call it equation $\overline{E}$— that had to be correct if Maxwell's wave theory was true. He then examined the experimental data and found that equation $\overline{E}$ did not accurately describe what happened. Thus, for Einstein, every rebuttal based on an appeal to Maxwell served as a reminder that Maxwell's waves did not work in this case.

Well then, if Maxwell was wrong, why did other physicists stick by him so fervently? Probably because Maxwell did work brilliantly everywhere else. Einstein had conceded that "the wave theory of light . . . has proved itself superbly in describing purely optical phenomena and will probably never be replaced by another theory." It was also true that Einstein's mathematical argument was novel and many physicists probably could not follow it in all its richness. But Lorentz, Planck, and Max Born could follow the math, yet they did not move on to light quanta.

Something beyond evidence and mathematics shaped this dispute.

Einstein was uniquely able to work in situations where language was ambiguous and theory offered no guidance. Most scientists, Thomas Kuhn reported, when forced to choose between a false theory or no theory will cling to the false one: "Though they may begin to lose faith and then to consider alternatives, they do not renounce the paradigm that has led them into crisis." Einstein was different. He rejected theories he knew to be false and went on working. While other people clung to Maxwell, Einstein pressed on like Columbus into the unknown ocean. Yes, he was a scientist and did not share the imagination of a poet who seeks language that will embrace reality's contradictions. He accepted the scientist's position that things are one way or another, but he was also like a star trapeze artist who leaps from one perch and flies beyond the net, confident that somewhere ahead is another swing.

To be blunt, Einstein had more courage than his colleagues, and his courage came from faith, although no honest preachers will claim Einstein's faith as their own. For Einstein did not believe in the Hebrew God, an immortal person who intervened in the affairs of nations. His faith was in the abstract proposition that the universe makes sense and runs on meaningful, physical law. If something is wrong with our current understanding we should immediately abandon that error in the sure and certain hope that a better understanding can be found. That great bravery mixed with grand ability is rare in every field. Shakespeare had it. Copernicus had it too, for he, like Einstein, first rejected the established picture of the cosmos and then cast about for a better explanation. Meanwhile their contemporaries could see the problems and recognize the need for something new but could not find the will to jump from their perches. However bumblingly and nervously they had come to it, the Nobel committee had finally honored Einstein for the fruits of his unusual mind and even more unusual courage.

Bohr immediately sent Einstein a congratulatory message, "The external recognition cannot mean anything to you. . . . For me it was the greatest honor and joy . . . that I should be considered for the award at the same time as you. I know how little I have deserved it, but I should like to say that I considered it a good fortune that your fundamental contribution in the special area in which I work [that is, the quantum theory of radiation] as well as contributions by Rutherford and Planck should be recognized before I was considered for such an honor."

Einstein replied, "I can say without exaggeration that [your letter] pleased me as much as the Nobel Prize. I find especially charming your fear that you might have received the award before me—that is typically Bohr-like. Your new investigations on the atom have accompanied me on the trip, and they have made my fondness for your mind even greater."

After their polite bowing, they made speeches that expressed mutually opposing views. One of Einstein's Tokyo talks was about "The Present Crisis in Theoretical Physics," in which he noted the battle over light waves versus light quanta and how both quanta and waves seemed absolutely necessary, although in different cases.

Six days later, in Stockholm, Bohr read his Nobel speech and in it he gave Einstein's light-quanta theory no quarter. His talk summarized the six lectures given the previous June in Göttingen, but at the Festspiel he criticized Einstein only implicitly. At Stockholm he was more direct. He began praising Einstein, noting that it was Einstein who first commented on the revolutionary importance of Planck's quantum discovery. Bohr then mentioned some important work Einstein did in 1907, while he was still a patent office clerk, in which Einstein had used the quantum $h\nu$ to calculate how much energy it takes to raise a substance's temperature. Einstein solved an old conundrum with this quantum equation and founded the science now called solid-state physics. Oddly, Bohr said Einstein's successful work on heat led to the formulation of the "so-called hypothesis of light-quanta," but surely Bohr knew that Einstein's light quanta had come two years earlier, in 1905. Or maybe he did not know. Bohr had been a university student in those years and had other things on his mind. Either way, Bohr's account allowed him to tell Einstein's story with a consistent flow—all the praise came up front and was followed by growing doubts about what Einstein had wrought. Bohr ended his summary of Einstein and quanta with the standard Millikan-style argument that although Einstein's equation described the photoelectric effect perfectly it was an impossible explanation. Bohr was not yet ready to leave his old perch on the trapeze.

Unknown to both men, something new had happened. Two days before Einstein's crisis lecture in Tokyo, an obscure American physicist named Arthur Compton read a paper in Chicago that would force the quantum experts in Europe to abandon their swings. But news did not fly in those years. Bohr gave his talk in Stockholm unaware of what had happened in Chicago. Einstein continued on in Japan. Compton's discovery was like a bullet that had been fired but has not yet hit its target. That collision would come in 1923.

In Japan, Einstein enjoyed a triumphal tour that outdid even his huge successes in America and England. He conversed with Japan's empress in French and was cheered by sold-out audiences who listened to him give scientific talks in German. His image filled newspapers. He played the violin, gazed on Mount Fuji, and attended kabuki theater.

There were also women. Folk wisdom reports: *Your husband runs around with many other women? That's not good. Your husband runs around with one other woman? That's bad.* Elsa had known from the start that her marriage to her cousin would not be so good. Her philandering husband would leave her at home to go off to an evening with somebody younger, prettier, and after he became world famous, what had been easy became even easier. The word *groupie* had yet to be coined, but young women who wanted to have sex with famous strangers were already common. Einstein was not as popular that way as, say, Babe Ruth, but neither was he ignored.

By the Japanese trip's end, matters had gotten bad. Elsa's oldest daughter, Ilse, had served as Einstein's secretary, but after Ilse married, Einstein hired a young stranger named Bette Neumann. Trite though it is to fall for the secretary, Einstein fell. It was the old pattern. As quantum physics grew increasingly frustrating and baffling, the emotional appeal of a new woman grew stronger. By the end of the Japanese tour, Elsa had something serious to worry about.

Just before the new year Einstein boarded another steamer for the voyage home. He did stop off for a month in Palestine, where he got to know a variety of "tribal companions" who were present in every form. The British governor was an intellectual, politically savvy Jew; Einstein also saw kibbutzim run by secular, socialist Jews after his own heart; he stood before the Wailing Wall, dismayed by "dull minded" Jews rocking in prayer: "A pitiful sight of men with a past but without a future." He helped lay the cornerstone of Jerusalem's Hebrew University, and then he was off to tour Spain.

By mid-March he and Elsa were back in Berlin. During their five-and-a-half month absence, the hyperinflation burning through Germany's money had destroyed the middle class and began pressing it beyond ruination to a kind of *über*ruin unknown to previous history. The bourgeoisie living on savings and fixed pensions were stripped bare. Elsa had had some money; now it was gone. Retired generals and diplomats who had spent their careers serving imperial Germany and who thought of themselves as distinguished people took to rooting through garbage cans for food. The only wealth that had survived was property and foreign currency, and even property became a loss the instant it was sold for cash. No matter how great a sum the property had brought, inflation at once set fire to the earnings and, within days, reduced them to ashes. Property was something to hold on to, if you could, or something to borrow money on, if you could find a lender, since the debt, whatever the size, would soon become meaningless. Einstein had no property, but his Japanese trip had provided him with a bushel of foreign currency so that neither he and Elsa, nor his two stepdaughters risked starving.

11

A Completely New Lesson

openhagen was snow covered when Bohr received a warning from Arnold Sommerfeld that Compton's bullet was on the way. Einstein was lingering in the warm zone, sailing toward Palestine, taking the pleasures of sunning on deck, quite oblivious to the teaser sent to Bohr. Sommerfeld had managed to escape Germany's inflation by accepting a yearlong visiting professorship in Madison, Wisconsin, and on January 21, 1923, he sent Bohr a heads-up that was a marvel of high excitement and low information, "The most interesting thing that I have experienced scientifically in America . . . is a work of Arthur Compton in St. Louis. After it, the wave theory of Rontgen-rays [X-rays] will become invalid . . . I do not know if I should mention his results [which had not yet appeared in print]. I only want to call your attention to the fact that eventually we may expect a completely fundamental and new lesson."

That should have kept Bohr pacing. The wave theory of X-rays dead? But then the wave theory of light, too, would be invalid and Bohr's whole contribution to physics might come under question. His picture of the atom depended on waves. During the 10 years since he first proposed that electrons change orbits while emitting light waves, physicists had found many subtleties in atomic spectra, but always the discoveries were interpreted as alterations in a light wave. Sommerfeld had been among the most fertile of these interpreters, and now he was reporting that the jig was up! How? Why?

Compton's idea was not published for several more months, so Bohr could only stew. In the spring when Compton's idea finally appeared, in an oddity of history, it was signed by the Dutch physicist, Peter Debye. Einstein was back in Berlin by then. Debye's paper hailed him as a prophet and his light quanta as real. The data cited, however, were not Debye's but Compton's.

The unknown Arthur Holly Compton was an American Mid-westerner with a striking, great head, broader even than Einstein's, that sat on a tall, wide-shouldered body. He looked like a bull, although temperamentally he was a Ferdinand, more apt to smell the flowers than charge a red cape. Scientifically, he was an experimentalist, like Millikan, out to uncover facts rather than joining the theoreticians who look for underlying principles. Also like Millikan, he had dis-missed Einstein's light quanta out of hand.

Compton spent years studying X-rays and the photoelectric effect without ever considering a role for light quanta in his experiments. He took Maxwell's wave theory to be as settled as the Copernican view that the earth orbits the sun. In 1919 he arrived in Cambridge to do a year of research at the Cavendish laboratory. A year's work in Europe was a standard way to round out an American scientist's edu-cation. Rutherford, godfather to Bohr's quantum atom, directed work at Cambridge, yet when Compton left he was still fully committed to classical theories of electric fields.

Compton studied X-rays, a new kind of radiation that had been discovered only at the end of the previous century. X-rays had seized the public's imagination because they could travel through closed drawers and sealed envelopes. The rays fogged photographic plates that lay hidden in such closures and the ghostly images they created pro-vided a visible sign of an invisible world. Throughout the years before the Great War physicists disputed just what X-rays might be. Some thought they were small particles, others argued that they were elec-tromagnetic waves, but by the time Compton received his doctorate in 1916 the answer was in: X-rays are akin to light, but they have a much higher frequency. They were part of the same electromagnetic spectrum that includes infrared, ultraviolet, radio, and all those other forms of invisible radiation that secretly flood our universe. This point had been settled by Einstein's friend, Max von Laue, and Sir Lawrence Bragg. Working independently, both showed that crystals could produce interference patterns in the same way that more coarse objects can interfere with light waves. So when Bohr read Sommerfeld's news that "the wave theory of Rontgen rays will become invalid," he knew that more than X-rays had come into question. The wave theory of light rested on exactly the same evidence—interference producing darkness in light—as did X-rays.

For years Compton had fired X-rays against a metal surface and checked the results. More than Einstein's meditating ever could, Compton's lab work matched the scientist's popular image. While Einstein sat in his study jotting ideas in a notebook, Compton bent over a big, noisy machine that produced zapping sounds and sparks. When X-rays hit human bodies they go right through the skin, but they scatter in all directions when they hit metal. When Compton began his studies, very little was known about X-ray scattering, and he was "merely" trying to learn as much as he could about the topic. What happens when X-rays cannot penetrate something? Where do they go? How does the metal react? Compton had no theories or hypotheses to test. He later wrote, "The mistaken notion is to get some idea and then try to prove it. . . . The real thing a scientist tries to do when he is faced with a phenomenon is to attempt to understand it. To do that he tries all the possible answers he can think of to see which one of them works best." This is science in the spirit of Galileo, who looked through a telescope simply to see what he could see.

At first glance, Compton's research project looked straightforward. X-rays were expected to hit metal in much the same way that waves from the Pacific Ocean hit a California cliff. They splash, eroding the wall and sending water droplets flying in all directions. There is also an invisible energy transfer from wave to rock, heating the rock a bit, as though the ocean waves were a dim form of sunlight. Most of the variation in all this splashing and erosion depends on the nature of the cliff rock, but clear as this analogy is, it was only partly helpful in understanding Compton's observations. He learned quickly that it did not matter what metal he used, the scattering patterns looked alike. This surprise was like discovering that the Pacific Ocean splashes off sandstone and granite in exactly the same way. Then there were secondary effects to sort through. For one thing, when X-rays hit metal there is a notable photoelectric effect. You don't get electricity when the Pacific meets California.

Another challenge for Compton was figuring out what were X-ray scatterings and what were other effects. It was like trying to understand Niagara Falls without knowing that light can form a rainbow. You see the Falls and the cloud of spray. You also see a perpetual rainbow deep in the spray. Where does that come from? From the water? The cloud? The rock? Compton's lab work, in essence, tried to sort secondary, rainbow-type effects from the primary, splashing effects.

X-rays were thought to be like light, so in many ways Compton's studies were similar to watching what happens when light hits a metal surface. If the metal is dull, the light scatters in all directions. If it is highly polished, the metal becomes a mirror and reveals the image of the light source. Sometimes too, light striking metal can produce a "fluorescent" effect—the metal gives off its own light. Obviously, if you are trying to account for light scattering, you need to know which light is scattered and which is added through fluorescent effects.

Compton patiently worked through all these complications, determining what was scatter, what was fluorescence, and what was simple illusion. He studied X-ray effects as he worked on his doctorate at Princeton and kept it up through work in Minneapolis, Pittsburgh, Cambridge, and then in St. Louis, where he taught physics at Washington University. Washington was a small and remote school, but it had the X-ray equipment Compton needed and did not have the established physicists whose thought would dominate their department. In St. Louis, Compton said, he was free "to develop what I had conceived of as my own contribution."

Neither Bohr nor Einstein had much patience for lab work. Experiments demand an ability to tinker with equipment and refine setups as the moment requires. It was more than asking questions and checking for answers. It meant sweat in the eye and grit on the thumb, whereas—if they had wanted—Einstein and Bohr could have done their physics while dressed in dinner jackets. One surprise Compton observed while wrestling with his X-ray equipment was that the scattered X-ray changes its frequency. He had explored the effect very deeply to make sure it really resulted from the scattering and not from one of the many secondary effects. Finally he had obtained a clear picture: X-rays approach the metal at one frequency and, when they scatter, they leave the metal at a different frequency. This change was like discovering that a violet flower's mirror image looks red. Compton, of course, could not know that this was exactly what poor Hendrik Kramers had wanted to look for in his proposed *experimentum crucis* and that Bohr had argued against. So Compton continued to puzzle over an explanation for his odd observation.

Compton eventually wrote a report for the National Research Council in which he set forth his X-ray data and observations. Although he still had no explanation for the frequency change, he included it and many other findings in his data, establishing numbers and equations that had long eluded physicists. He also showed a new experimental way of using X-rays to calculate the number of electrons in an atom. This technique, at last, gave physicists an independent technique to confirm quantum assumptions about atomic structure. Compton had moved to the first rank among experimental physicists, but he still wondered about that frequency change. Why does a scattered X-ray change frequency but light bouncing off a mirror does not?

Compton had moved into Einstein's sort of physics, the kind that calls for obsessive thinking in shower stalls and during walks along Washington University's tree-lined paths. Finally, as a new school year was getting under way and Einstein was setting off for Japan, Compton took his great step. Although Maxwell's wave theory offered no explanation for the change in X-ray frequencies, the light-quanta hypothesis did. Kramers' lost idea was back. If light quanta exist, they will inevitably lose some energy during a collision with an atom. In wave theory, frequency and energy are not tied together, but in light quanta they are bound together in mathematically tight handcuffs. Lose energy, lower frequency. The math even explains why the change is detectable in X-rays but not mirrors. Visible-light frequencies are already low, so a small drop in energy will produce a very small change in its frequency. Red light, that is, stays red. X-rays, however, with their high frequency will produce a greater frequency shift per unit of energy lost.

In November, Compton gave his breakthrough lecture, telling his students that the theory of light quanta was indispensable for explaining his X-ray scattering experiments. Centuries earlier Isaac Newton had presented his discoveries about light to bored and bewildered students who had no feeling for the revolution being set before them. The same story applied to Compton. He had a bullet aimed for the heart of classical physics, but few appreciated it. In December he reported his discovery to the American Physical Society's annual meeting, and two weeks later he sent the report to *The Physical Review*. The journal, however, was not in the habit of receiving revolutionary papers and did not schedule its publication until May 1923.

Meanwhile, in Zürich, Peter Debye read Compton's report to the National Research Council. Like Kramers, Debye had already considered the possibility of change in frequency, but had never published it. He might not have thought that the experiment was worth the trouble because it would only prove what everybody, except Einstein, already thought: light quanta do not exist. But after he read Compton's paper, Debye knew what the data meant. In March, while Compton's paper was undergoing a leisurely preparation for publication, Debye's paper arrived at the office of the *Zeitschrift für Physik*, the world's leading journal for news of the quantum revolution. In April, still a month before Compton's article appeared, Debye's article was published. Einstein, Planck, Bohr, Heisenberg—every physicist interested in quanta—now knew that the wave theory of electromagnetism was down, possibly (even probably) out.

Usually when two scientists discover the same law, the more celebrated of the two gets the larger slice of fame pie. Newton was credited as the father of the calculus, even though Leibnitz published first and used the better notation system. As the father of gravity, Newton already well outdistanced Leibnitz in public acclaim. Similarly, Darwin was also well-established as a scientist and popular author when Alfred

Wallace submitted his essay on natural selection. Darwin's fame grew spectacularly, even though Wallace's paper still stands as a model for a clear and direct statement of how natural selection works to produce new species.

In April, Arthur Compton looked doomed to a similar fate. Debye had gotten into print first and was known to all of Europe's important physicists. He had been at Göttingen and was now in Zürich. Compton was off in wild-west country. It was Sommerfeld who saved Compton's fame. Months before the Debye article appeared, Sommerfeld had already begun lecturing in America about the importance of Compton's work. Furthermore, Sommerfeld was the author of the leading textbook on quantum theory and, in the summer of 1923, as he revised his book for its latest edition he described a "Compton effect." So Debye was consigned to the role of Alfred Wallace. To his credit, Debye took his fate without bitterness. Many years later a historian asked him if the discovery should be called the Compton-Debye effect. No, Debye insisted, the person who did most of the work should get the name.

Another puzzle is why Einstein never anticipated the Compton effect. If, using theoretical principles alone, Kramers and Debye could predict the frequency change, why didn't Einstein think of it too? Perhaps he did. Einstein's letters often included obscure references to ideas that he had chased down as full of promise and later rejected as temptations from the devil. Einstein had a tough definition of understanding. It would not have been enough for him to say, *Gee, if the energy decreases, then v will have to decrease too.* Before proposing any such idea he would have insisted on knowing how much the frequency decreases. He had provided exact numbers for the way gravity would bend starlight. A prediction of the Compton effect would have had to be as precise. Compton's equation linked energy change and frequency change precisely with the angle of scattering. Different angles led to different numbers. Einstein would have demanded just that precise an equation before he reported his idea, and perhaps fame had become just too distracting for him to do the math.

There was also a dog that did not bark. Johannes Stark was silent through all this, even though back in 1909 when he and Einstein stood alone, Stark had called for experiments with X-rays and angular variation that, if successful, would prove light quanta were real. Stark had never quite predicted the Compton Effect, but he came close, close enough to tempt even a milder man to say "I told you so." Yet Stark kept silent. Apparently he so resented Einstein's success that he no longer cared to remind anyone of his good, ground-laying work in quantum theory.

Compton's stride settled most doubts. Scientists are not like judges who can make rulings and even allow a person's execution, when a law points one way and a fact points the other. Scientists must go with the facts. When Einstein had only logic on his side, his would always be a minority position, an embarrassment for the facts of light's interference but not their refutation. The Compton Effect put a new fact on the table. Electromagnetic reality can behave like a billiard ball bouncing off other billiard balls. Compton's step took him and all who followed him into the world that Einstein alone had inhabited for many years. From there they could see not only the meaning of hv as light quanta but could also grasp the problem that Einstein had been posing since 1909—light has the properties of both particles and waves, so what is it?

Physics had reached a crisis it had not known since the days when Johannes Kepler showed that the earth is one more planet making elliptical orbits around the sun. That put an end to Ptolemy's system, yet there was still a mass of Aristotelian theory that explained motion very well on a resting earth but made no sense if the earth itself was moving. It required Galileo's and Newton's combined efforts to create a new physics that overcame those contradictions. Now, in hv, physics had a mathematical idea as successful as the moving earth, but which made everything known from before seem as false as wartime news. On the bright side, this revolution was too important for Einstein to continue wasting his time being famous. Compton's bullet served rather like the ghost of Hamlet's father, who returned to whet the prince's "almost blunted purpose," and Einstein resumed thinking full-time about the mysteries of physics.

12

Slaves to Time and Space

S een from a ferry, Copenhagen appeared first as a smoky cloud, then as factory chimneys poking above the horizon. Einstein was coming from Sweden where he had given his delayed Nobel lecture. The Swedes had cheered him, of course. He had spoken before an audience of 2,000 souls in the port city of Göteborg, which was celebrating its tercentenary jubilee.

Beneath the smoking chimneys, as Copenhagen's green-copper domes and red roofs rose into view, Bohr was waiting for Einstein to land in Denmark. As the ferry moved into the harbor, Bohr was still refusing to follow Compton's stride into the world of light quanta. Bohr dismissed Compton's effect as an illusion. His assistant, Kramers, compared the theory of light quanta with "medicine which will cause the disease to vanish but kill the patient." He meant, perhaps, that the theory would explain the data but abolish coherence. Even so, Bohr found plenty to talk about when Einstein landed. Presumably they said Hello and maybe something about the brief voyage and perhaps something about Sweden, but very quickly they were deep into physics. They were lost to the world even before the tram delivered them to Bohr's home.

Regrettably, even though by that time Einstein was very well known and apt to be recognized wherever he went, no passengers aboard the tram have left an onlooker's account of how these two men conversed, or rather spewed speech at one another. Both were well known to be compulsive talkers; neither was distinguished as a compulsive listener. Einstein did not demand that the so-called listener actually comprehend what he was saying. A politely open pair of eyes was good enough. Elsa could be good enough. Bohr liked to turn listeners into secretaries, asking them to transcribe his words as he spoke. In neither case was audience participation required or looked for. Undoubtedly, therefore, as the two men spoke they sometimes stepped on each other's sentences and often interrupted with "Yes, buts" that soared off on tangents.

Bohr, sober and long-winded, still insisted that the light quanta idea was irrational and therefore false.

Einstein, direct and witty, thought contradictions in nature were mysteries to be explained.

Bohr saw the light quantum as more impossible than mysterious. X-rays were waves, period. The light quanta formula even included the v for a frequency and frequency could be measured only by an apparatus that treated the X-rays as waves.

Of course Einstein had recognized these puzzles when Bohr was still a college student. He knew too that the h was just as much a part of the quantum formula, a little, unshrinkable, unvibrating number.

The two men were pipe smokers, although it is a challenge to talk energetically and keep a pipe lit. It is even harder to light a pipe bowl while interrupting a fast-paced talker, so their smoking rituals would have forced each man from time to time to listen quietly to the other.

Bohr, too, was used to the quantum oddities. He knew that in classical theory a wave could shrink to something so small that it was just an infinitesimal burp away from being nothing at all, but energy that weak cannot support an electron in its orbit, just as wings cannot keep a too-slow airplane in the sky. Bohr did not look for any physical reality behind his symbols; h was just a number and hv was just a combination of numbers that was not to be taken as something's name.

Einstein did look for a symbol's meaning and once used the metaphor of a beer stein in a statistical analysis to prove that beer never comes as a free fluid but is always packaged in steins. Bohr was never impressed by either the humor or the quantitative abstraction behind the beer-stein metaphor. He held that although beer was served only in multiples of pint-sizes (steins of a pint, quart, half gallon, and so on.), beer was still only a fluid and never a mug.

Looking outside the tram Bohr noticed that they had been conversing so intensely they had missed their stop, and not by just a block or two. The two men hurried to the exit and made their way across the tram tracks to take another ride back. Bohr was the younger man and still had a full head of hair, but Einstein was the more imposing—taller, stronger, and he sported that great head. If our regrettably silent passenger had been on the tram and had watched the two men go, Einstein would likely have been the one to seize the stranger's eye.

Copenhagen's port district was busy, but away from its commercial heart the city quickly felt snug and provincial. The two physicists were not surrounded by a Berlin-like metropolitan vigor as they waited for an electric tram. Bicyclists were sprinkled all over Copenhagen streets.

For years, Bohr had been dubious about the conservation of energy at the quantum level. Now the Compton effect was forcing a crisis. Compton and Debye's analyses worked only if energy conservation held true.

Einstein saw no reason to try anything as radical as rejecting the conservation of energy.

Bohr considered rejecting energy conservation to be far less radical than marrying particles to waves.

Einstein disagreed, believing that physical sense itself depended on keeping energy conservation.

The issue seems abstract, remote from anyone's life, but the problem was as fundamental as the need for air. How shall we understand our lives? Lions and lambs seem never to worry about that, but humans always do. Judges anguish over how to apply the law. Believers kneeling in front of a candle grapple with some private hope or fear. An artist twisting colors on a pallet looks fiercely for the mixture that is true to the moment. Bohr's and Einstein's concerns were just as personal and just as universal.

Bohr's heart grasped things metaphorically, seeing abstractions as tools that distracted from the particular.

Einstein's soul relied on abstractions that clarified concrete experience's ambiguities.

Yet both men were physicists, great physicists, and even in their seemingly infinite disagreement they shared some common ground. Einstein published a paper that summer, one he had written with his chum Paul Ehrenfest, which Bohr thought correct. Compton had suggested that light quanta ricochet like billiard balls and jump from one frequency to another. Einstein, Ehrenfest, and Bohr, too, agreed that when the incoming X-rays hit an electron it was absorbed. The electron emitted new light at a lower frequency. They disagreed over how and when the new X-ray was emitted.

For Bohr the emission was unrelated in either space or time to the absorbed X-ray.

Einstein believed that the absorption and emission was instantaneous.

Bohr could not avoid one weakness in his position. He had instinctively rejected the Compton effect, but he had no better way to explain Compton's data. All he could do was voice his doubt.

Einstein knew his own weakness too. He still had no good idea how to synthesize waves and particles into one comprehensible, concrete thing.

Matchboxes in Denmark had the king's portrait on them. Christian X looked on quietly whenever Bohr relit his bowl.

Oh my God! They had missed their stop again. The tram was approaching their starting point. There was not much to do beyond exchanging sheepish grins, dismounting their public conveyance, crossing the tracks, and trying again. The two men stood at the tram stop, the very picture of absurd genius. Many would have said they were the two smartest men in the world, but they stood side by side at an anonymous way station, yo-yoing back and forth across Copenhagen because they were too distracted by their own intelligence to find their way home.

Of course there was more at stake in their endless talk than truth, physics, and nature's way. Both men were risking their reputations and achievements. Bohr's research since 1913 had concentrated on atomic structure. He held that $h\nu$ limited electron orbits, and that to get from one possible orbit to another the electrons "jumps" from one place to another without going through space or time.

Einstein's ambition was to unite his general theory of relativity with the atom's electromagnetism. For that scheme to work, he was going to have to express the unity in terms of space and time.

Bohr did not like Einstein's effort at a spatial mapping. If light quanta can bang into electrons like meteors colliding into planets, space and time stay firmly in the story.

Einstein was pleased with the way the Compton effect's X-rays scatter in a positively Newtonian manner that allows for measuring scatter angles, energy exchanges, and momentum transfers without ever having to perform a "jump."

As it moved, the tram passed bicyclists, but the cyclists sometimes caught up with the tram again when it paused to take on and release riders.

By arguing against energy conservation, Bohr defended his own contribution. Bohr's orbiting electrons could be mapped in space and time, of course, but space posed no limits on how "far" an electron could jump when it shifted from one fixed orbit to another, and no clock determined when it was time to make a jump. These jumps just happened. The electron was here, now it is there; it did not travel between here and there; it just changed. Instead of a calculus of continuous changes, the ultimate quantum theory would have to predict the discontinuous states that electrons "jumped" to.

Perhaps Bohr's success was why Einstein never predicted the Compton effect. Bohr's electron did not use space and time in a classical way. Einstein was skeptical about the Bohr atom, but skeptical physicists had for years been making progress by accepting jumps and ignoring the continuity of space-time. Maybe Einstein had finally lost his nerve, like a poker player who thinks his full house looks good but wonders why the other fellow keeps raising. Maybe Einstein had seen the whole thing—light quanta scattering, their energy changing, the angle of scatter providing the decisive link between changes in frequency—and yet he had hesitated. It would be terrible to announce a decisive experiment and then have it fail, not because there was something wrong with light quanta, but because space-time does not hold at subatomic distances.

Bohr, too, had been growing nervous. His theory's initial success had, in recent years, become increasingly problematic. Despite his explanation of chemistry's periodic table of the elements, Bohr's theory had yet to work in all its details for any atom more complicated than those with a lone electron.

If Einstein, usually so cocky, had grown nervous, the Compton effect restored him to full confidence. His defiant insistence on light quanta's reality had finally been rewarded.

Meanwhile Bohr's long perch on a pedestal had grown precarious. The Compton effect's success at describing continuous change brought the whole discontinuous structure of the atom into crisis.

So Bohr and Einstein both had their excuses for riding back and forth through Copenhagen on the city's tram system while puffing their pipes. Both were slaves to time and space, but they disagreed profoundly over how tightly space-time bound the atom.

13

Where All Weaker Imaginations Wither

When science was disarrayed, Einstein was disarrayed, but the Compton effect had restored sense to science. Quanta worked as Einstein had imagined and quantum radiation was no longer quite so terrible a blank. It was difficult, mysterious, solvable. Einstein had known the prophet's struggle before—in his desperate days in Bern as an unemployable physicist his statistical analyses drew no support until he used the technique to understand Brownian motion; then again, even after other physicists hailed him as a wonder he still could not persuade most physicists to take the contradictions between relativity and gravity seriously.

By December 1923, whatever uncertainty Einstein might have felt about the relation between quanta and space-time had passed. It was time to put distractions of fame and life's other detours aside. Germany, Zionism, and love persisted, but he seemed determined to keep his eye on science. He made a public appearance at the Prussian Academy to discuss before the assembled savants an approach to the quantum crisis based on space and time. He reminded his audience that wave theory still explained much about light. Now that his colleagues were all nodding their heads in agreement that the Compton effect proved light's particle nature, Einstein was at the front insisting that light waves still worked too. A replacement for both waves and particles was necessary and, he said, he was looking for that replacement in mathematics where, he expected, space and time, particle and wave could be joined in a new law just as Maxwell had joined electricity and magnetism. Einstein was looking, he said, for a quantum field that would be based on his earlier work on relativity.

Einstein's return to the public's eye reflected a recovered civic boldness. In Munich, the month before Einstein presented his paper, the unknown Adolph Hitler had joined with the war hero General Ludendorff to seize the Bavarian government. Hitler's action combined audacity and absurdity in an amalgam that would characterize his behavior straight through to his suicide. General Gastov von Kahr, Bavaria's prime minister, was scheduled to make a speech at a beer hall. Rumor had it that Kahr planned to announce the restoration of the Bavarian monarchy, something unacceptable to both democrats and national extremists. Hitler's brown-shirted hoodlums surrounded the beer hall and seized General Kahr. Hitler then proclaimed a new government with himself as its leader. He dreamed that the masses would rally to his side, but that nonsense came to naught. Hitler's action, however, did cause much excitement and rumor. Nobody was sure at first how large Hitler's plot was or how widespread the coup was. Sensible people tended to suppose that Hitler's action could not really have been as crazy and hopeless as it looked. Planck was especially worried for Einstein, fearing that he might be seized or arrested as part of a wider plot.

Nobody could see it clearly at the time, but Hitler's absurd theater at a Munich beer hall marked postwar Germany's rock bottom. Brutality was no longer in charge. Then, the next month, the government replaced the ravaged currency with a new money called the Rentenmark. One Rentenmark was worth a trillion (sic!) old paper marks. Just producing all those ruinous trillions had taxed Germany's production capacity. Hundreds of paper mills and thousands of printing presses, including even newspaper presses, had been diverted from useful activities to supporting runaway inflation. Would the new money work? Nobody could be certain. When Einstein made his presentation on quanta and space-time, people were just beginning to hope that a new, more law-based Germany might have been born.

In the former enemy countries, many ordinary people still wished Germany ill. The war was five years gone, but the hatreds were alive and active. Scientists like to picture themselves as above the emotional moods that hold the masses, but of course they are not. German physicists were still scorned by non-German scientists. The Solvay Conference brouhaha had resumed. Another meeting was planned for 1924. Einstein's instincts said attend, especially when there was something like the Compton effect to explore, but again Einstein was to be the only German participant, and again his fellow German physicists insisted that he should not attend. Before Compton's bullet, when the world was rich in distractions, Einstein might have easily skipped Solvay. But now it was clear that light quanta worked excellently in a scheme of space-time. He was ready to concentrate on science and forget politics. Others, however, took nationality seriously, and Einstein admitted to Madame Curie that "the disinclination of Belgians and French to meet Germans [is] not psychologically incomprehensible to me," but it was complicated. In Germany his fellow scientists expected him to be a good German and to refuse any honors that were forbidden other Germans. Meanwhile many rightist German politicians and agitators were eager to put Einstein among the first Jews to be driven off.

Whenever one side was ready to forget the war, there was somebody on the other side who remembered. Abram Joffe, a Russian physicist, came to Germany to renew old ties. During the century's early years he had studied in Munich and had known Philipp Lenard. Joffe looked up his old companion, but when he arrived at the University of Heidelberg, Lenard sent the porter to say he was too busy to speak with "the enemies of his fatherland."

This continuing bitterness was a great help to Bohr and his institution in neutral Copenhagen. Bright young German students, welcome almost nowhere else, could come there and mix with the up and coming stars of other countries. In December, while Einstein was speaking before the Prussian Academy, Bohr and Kramers were having excited discussions with a recently arrived American. His name was John Slater and he had given Bohr an idea about how to defeat the Compton effect.

Yet Germany was still the capital of quantum physics. In troubled Bavaria, Arnold Sommerfeld turned out promising students one after another. "You have produced so much young talent," Einstein marveled to Sommerfeld, "like stamping them out of the ground." And in Göttingen, where Max Born had landed a star position, student physicists received their final polish. Einstein in Berlin with the friendly Maxes—Planck and von Laue—found this tension over Germany profoundly distracting. He was trying simply to concentrate on science and draw new insights from the physics of $h\nu$, but the German question persisted.

Einstein wrote to Lorentz, "Sommerfeld believes it is not right for me to take part in the Solvay Congress because my German colleagues are excluded. In my opinion it is not right to bring politics into science matters, nor should individuals be held responsible for the government of the country to which they belong. If I took part in the congress I would by implication be an accomplice to an action which I considered most strongly to be distressingly unjust." He closed by pleading with Lorentz, "I would be grateful if you would send me no further invitations to the Councils. I hope not to be put in a position where I am obliged to refuse an invitation, as such a gesture might hinder any progress toward reestablishing amicable collaborations between physicists of all nationalities."

He sent Madame Curie a splendidly aristocratic yelp, "It is unworthy of cultured men to treat one another in this type of superficial way, as though they were members of the common herd being led by mass suggestion." He returned to his equations, working in his upstairs study with Newton's portrait on the wall. He mainly sat quietly with a pen and paper, playing with images and contemplating what they meant, trying out his notions with mathematical symbols.

Thinking about almost any physics principle, Einstein told Count Kessler that year, almost always led to progress because, without exception, scientific propositions are wrong. Every generalization about nature leads to contradictions. Einstein, sitting quietly in his chair, lost to the world, following thoughts that were inaccessible to all others, visualized ideas, looked for contradictions between rules and facts. When he found a discrepancy he looked for ways to restate the rule so that the facts survived while the contradictions were resolved. When Einstein wrote equations on paper he was either testing an idea—does $a+b$ work in this case?—or, trying to resolve a problem he had noticed,—why doesn't $a+b$ work this time?

Niels Bohr, about 200 miles northwest of Berlin, thought many classical descriptions had it pretty much right. Instead of tinkering with equations, he looked for an analogy that applied classical principles to novel observations.

Einstein's challenge was that, after years of looking, he still lacked a successful understanding that linked waves and particles or predicted the direction that light quanta will travel when emitted "spontaneously."

Bohr's trouble was that quanta did not fit into any analogies he knew.

Einstein planned to keep looking for another new idea; Bohr's plan was to resist the particle evidence as long as possible.

But the accursed German question kept getting in Einstein's way. He finally ended one recurring hullabaloo by ending the ambiguity of his situation. On February 7, 1924, Einstein issued a statement that 10 years earlier, when he joined the Prussian Academy, he had acquired the rights of Prussian citizenship. It was a nice way to put it—talking about rights rather than duties. After rejecting Solvay and embracing Prussia, his commitment to democratic Germany again looked fixed.

Two months later Einstein published a newspaper article in the *Berliner Tageblatt* on "The Compton Experiment" in which he described the crisis over radiation's particle and wave properties. At the same time, Bohr published a paper, jointly attributed to Kramers and the American student Slater, denying that the crisis had to persist. Slater had given Bohr the idea he had lacked the previous summer. Bohr now thought he saw a way to account for Compton's data without using light quanta.

Compton's interpretation of his data had assumed that all energy and momentum is conserved. No energy is destroyed; none is created. The same held for momentum. Bohr now thought there was a way around that idea. In Copenhagen, during an introductory bull session when Slater talked about what was on his mind, he spoke vaguely of a virtual field guiding the light quanta. It was the old, tantalizing but unfruitful idea of the ghost whip guiding wave-particles, but Bohr and Kramers suddenly saw a way to adapt this image to their needs. Instead of guiding (as Slater imagined), the field could communicate. Instead of concerning quanta (as Slater proposed), the field could concern itself with electrons. Slater protested, but Bohr and Kramers talked to him. They talked some more. They talked long and hard. Slater was a bright young American, a whiz kid who knew himself to be lucky to be in Copenhagen. Niels Bohr even took him seriously. On the other hand, weren't they changing his idea rather drastically? Bohr talked some more and in the end Slater agreed to add his name to a paper that transformed his notion that a field guided the light quanta into a field that provided instantaneous communication between electrons. The paper with this idea was titled "The Quantum Theory of Radiation" but was more commonly know by its authors' last names, Bohr-Kramers-Slater. Even more commonly it was referred to by the authors' initials, BKS.

The field in BKS was called a "virtual" field, meaning it did not really exist. A real field, a magnetic field for example, includes the energy needed to swing a compass needle. The BKS virtual field had no detectable energy. Yet the BKS field did have some real effects. The field's electrons all communicated instantly with one another. Some electrons would absorb energy, others would emit energy. These emissions might have appeared to Compton as scattering, but they were just random radiation emissions. Energy was not being conserved moment by moment through every action. It was conserved over time and across the whole virtual field. BKS was extremely vague about how this happened. The paper had exactly one equation, and that one was Bohr's famous quantum lines equation from 1913. Mathematics did nothing to further or clarify the BKS argument.

Einstein was not impressed and did not see BKS as a threat to his own work. More distracting was the contradiction people saw between his Zionism and his commitment to Germany. Nationalists considered his Zionism as barring his claims to be German while many Jews believed that assimilation into German culture prevented any interest in Zionism. Einstein handled the questions the same way he did the particle-wave contradiction, by underlining both. He was thoroughly secular, with no sympathy for religious tradition, but he joined Berlin's New Synagogue to reaffirm his solidarity with Germany's Jews. Thread by thread he was getting his life back in order so that there would be no distractions from his physics.

Even when he went out for an evening, physics held his thoughts. Janos Plesch, a fashionable doctor who treated and socialized with Einstein, described how his famous patient talked and laughed with Abram Joffe, the same Russian physicist whom Lenard had refused to see. The two were puzzling over the imponderables of physics on the quantum level and laughed gleefully at various absurd solutions. Einstein was back to having fun with his work.

The BKS paper, however, was not so much fun and Einstein did not hide his distaste. "What does Einstein think?" Bohr asked an up-and-coming student in Göttingen named Wolfgang Pauli.

The reply came back, "Completely negative."

Einstein wrote to Max Born and his wife, "I should not want to be forced into abandoning strict causality without defending it more strongly than I have so far. I find the idea quite intolerable that an electron exposed to radiation should choose *of its own free will*, not only its moment to jump off, but also its direction. In that case, I would rather be a cobbler, or even an employee in a gaming-house, than a physicist. Certainly my attempts to give tangible form to the quanta have foundered again and again, but I am far from giving up hope. And even if it never works there is always that consolation that this lack of success is entirely mine."

Bohr had stung Einstein in BKS by using one of Einstein's great ideas. In 1917 Einstein had shown how to deduce the quantum's existence by assuming the conservation of energy and momentum. Einstein had always been troubled by that paper because it offered no way to predict when or in what direction a light quantum would be emitted. That, for Einstein, was an issue yet to be resolved. Now Bohr, citing Einstein, was using random quanta emissions as an argument against the conservation that Einstein had assumed in the first place.

Einstein was still grumbling about BKS's irrationality in June when he visited the university at Göttingen. It was a memorable visit for

Werner Heisenberg, who had become Max Born's assistant. During Einstein's visit, Heisenberg was introduced to the great man and invited to join him in a private stroll. Heisenberg was even more mathematically oriented than Einstein and so seemed unlikely to be much impressed by the mathematically empty BKS theory. He had told his good friend Pauli, "I do not see [BKS] as an essential progress." But then he went to Copenhagen, where Bohr gave him the full treatment, grinding Heisenberg into a BKS enthusiast. On his walk he told Einstein as much.

Einstein had a long list of objections. He had thought about this ghost field idea many times over many years and concluded it was not the answer. Energy conservation had been central to whatever success the study of radiation and thermodynamics had had. First, it had been critical to establishing the conditions that led to Planck's discovery of the quantum. Then it had been the fundamental assumption in Einstein's own argument that light quanta were real. Nature seemed strictly to obey the conservation laws. Now Bohr was proposing a virtual field with action at a distance. All right, but why should we assume that nature was ready to abandon conservation when it allowed actions at a distance?

We can only assume that Heisenberg had little to say in defense. He was bolder, brassier than the average German student, but he probably was not cavalier enough to challenge the world's most famous scientist while he defended a cornerstone of physics.

Einstein also disliked BKS because he had no wish to abandon causality. For him causality meant lawfulness. Nature does not do just anything it damn pleases. Certainly Einstein was not yet ready to abandon the assumption that had supported all his previous efforts in the field.

Again, Heisenberg was probably silent, or in un-hunh mode. He too had been trained according to conservation and causal assumptions. What could he say?

Einstein had technical arguments as well. If it were not for the laws of conservation, engineers could build perpetual motion machines. Einstein proceeded to describe a BKS perpetual motion machine: place a sealed box containing radiation in empty space, where outside radiation cannot get at it. The box would be knocked about by the internal radiation, and this knocking would increase, becoming ever more ferocious as the radiation inside the box continued. Without conservation to counter the increase, the box would bang about with forever increasing violence.

Heisenberg might have had more to ask and say about arguments like that one, but as Einstein's physics was perfectly proper he could not have argued for long.

Another of Einstein's technical objections considered classical optics. Bohr had been so busy hunting for a way to explain Compton scattering that he had ignored the long-studied, ordinary scattering such as what happens when light bounces off a mirror. BKS was miserable at explaining that familiar event. In a world where BKS held true, the phrase "mirror image" would refer to a blurry, off-color resemblance.

Einstein "had a hundred arguments" against BKS, Heisenberg gloomily reported the next day, but the striking fact is that Einstein's talk did not end the dispute. Heisenberg stayed with BKS and refused to recognize Compton's scattering data as the chit that authorized light quanta's reality.

There is in the story of Bohr and Einstein a mysterious element that seems irrecoverable: the force of Bohr's personality which, like Eden's fruit, once tasted could not be forgotten. "You can talk about people like Buddha, Jesus, Moses, Confucius," a student recalled, "but the thing that convinced me that such people existed were the conversations with Bohr." With Bohr long dead, that personality works today the way sex works in stories read by bright 10-year-olds. We can only gather in some vague way that Bohr's presence and persuasiveness was powerful, but we cannot really grasp what it was about Bohr that gave him such authority among physicists. It was the kind of aura that Einstein had among nonphysicists—the force of an oracle. Such a thing had not been known among physical scientists since Giordano Bruno and doubtless many people had supposed that a scientific oracle would never be known again.

The public admired Einstein and accepted him as a genius despite their inability to understand him, let alone defend his ideas—perhaps even because of that inability. Physicists, especially young ones, even more especially young ones like Heisenberg who had met Bohr, were inclined to credit Bohr's instincts, even though they could not defend them. Physics had become so abstract—and Einstein's light quanta threatened to make it even more abstract—that Bohr's confidence that it all worked, even if the working sounded a bit obscure, was persuasive.

Older physicists, like Einstein and Planck, who knew what it was like to have a terribly perplexing mystery click into place, were less given to embracing obscure prognostications. Even when they liked the idea that the conservation laws were merely statistical, older scientists balked at BKS's philosophical obscurities. Erwin Schrödinger, for example, was a 37-year-old professor of physics in Zürich who had given his inaugural lecture on chance's role in physical laws. He immediately sent Bohr a congratulatory letter, praising him for daring to resist the quantum revolution with such a "far reaching return to classical theory." There was just one little obscurity that Schrödinger had trouble with. Exactly what did Bohr mean by the word *virtual*? To Schrödinger, something that produces physical results is real, not virtual. How poetic, Schrödinger wondered, was Bohr being?

But trying to pin Bohr down on a word's meaning was like trying to cage a beautiful cloud. He could no more be cross-examined on such a point than the Oracle of Delphi.

Einstein was confident. He wrote to Ehrenfest that Bohr, Kramers, and Slater had sought to abolish free quanta but added that free quanta "would not allow themselves to be dispensed with." The coming revolution was not to be denied. Einstein had just one other loose thread from his personal life to snip off. He had resolved the ambiguities about his citizenship and ethnicity. He was a German Jew. Now he had only the confusion over his marriage to settle. He sent his secretary, Bette Neumann, a note telling her that he had to seek a happiness in the stars that was denied him on earth. One hopes she saw it for the caddish, self-centered poppycock it was. Where was she supposed to seek happiness? Einstein's Saint Francis side was again taking control over the Pantagruel. His emotions had no more time for women. He was returning his emotions full-time to science, not to Elsa.

Meanwhile, the physics world was growing restless at this dispute between Bohr and Einstein. Wolfgang Pauli probably spoke for the majority when he told Bohr, "The available data are not sufficient to decide for or against your view. . . . Even if it were psychologically possible for me to form a scientific opinion on the grounds of some sort of belief in authority (which is not the case, however, as you know), this would be logically impossible (at least in this case) since here the opinion of two authorities are so very contradictory."

While ordinary physicists wrestled over who was right, Einstein had moved on ahead. Correspondents around the world—unknown thinkers like Louis de Broglie in Paris and even somebody from Dacca, India named Saryendra Bose—had given him new ideas and he was out ahead again, thinking about parts of physics that nobody else was minding. When Fritz Haber wrote to Einstein, "I was in Copenhagen and spoke with Herr Bohr," it was news from a battlefield that Einstein considered conquered. "How strange it is," Haber added, for he did not think the battle over, "that the two of you, in the field where all weaker imaginations and powers of judgment have long ago withered, alone have remained and now stand in such deep opposition."

14

A Triumph of Einstein Over Bohr

Berlin at the end of 1924 was no longer the desperate city that had chased out an emperor. Money had been good for a year now. New buildings were rising across the city. The enthusiastic fusion of art, revolution, and unreason had passed. An expressionist stage play even had a character say, "Let us hear no more about war, revolution and the salvation of the world!" The euphoria had gone, yet there was still optimism; there was art; there was achievement.

Einstein toured a monument to all three when he traveled to Potsdam, in Berlin's suburbs, to attend the opening of a new building known as "the Einstein Tower." It was a squat (five-story), concrete, lighthouse-styled structure with a solar observatory on top and an eye-catching entrance at the bottom. Einstein traveled with the architect, Erich Mendelsohn, a war veteran who had starting making drafts of an astronomy center while he huddled in the trenches. Mendelsohn led Einstein through the building, showing off both its design and its equipment. From the windows Einstein could see the woods, rivers, and fountains of Potsdam. The tower's supporters liked the way it combined form and fun. Its opponents disliked its modernity and lack of German flavor. Einstein kept his counsel. The builders had sponsored it to test Einstein's predictions about gravitational effects on light frequency. Einstein's tour might have persuaded him—correctly, as history would show—that the tower would be of only minor help in proving his theory. After tromping through the building in silence, he and Mendelsohn went to a meeting of the building committee. There, finally, Einstein whispered one modern adjective in Mendelsohn's ear, "Organic."

Modernity had become such a part of the new Germany that the quantum crisis, too, made its way into popular discussion. Science always has its buffs, amateurs who know what the disputes are, and by January 1925 German life had become orderly enough for science fans to notice a debate over nature's lawfulness. The question even made its way into the *Leipzig Illustrated News*, a stylish newspaper filled with photographs. Into its gaudy pages, between the pictures of carnivals and catastrophes, Wilhelm Wien, the one-time leader of physics' anti-Berlin movement, provided a story about the quantum debate. "The notion that nature is comprehensible," Wien wrote, "is identical with the conviction that all natural processes can be reduced to . . . invariably valid natural laws." BKS, Wien said, was trying to sidestep fixed law and replace it with a statistical foundation, and he insisted that statistics not based on a deeper law "will never be recognized by physics as something final."

Wien had gone to the heart of Einstein's alarm. When BKS broke with energy conservation, it broke with the idea that things happen for a natural reason. Statistical thinking was not new in physics. Einstein was a master of it. His explanation of Brownian motion was a perfect illustration of how to use statistics to take a meaningful step. Pollen floating on water is knocked to and fro by molecules that cannot be seen or measured directly, but by using statistical reasoning we can determine the number of molecules present. Einstein's statistics assumed there were underlying reasons for an action, but BKS statistics looked for no such deeper meaning. Indeed, its rejection of energy conservation did away with deeper causes. That was why Einstein told his friend Max Born he found it "intolerable" that an electron would be granted "free will" to act when and how it "chose."

Unexplained actions, of course, were a commonplace in physics. They were what kept physics incomplete and assured physicists of future employment. As long as there were unexplained phenomena there would be a need for scientists to try to understand them. Without that search for explanations, Archimedes could get back in his tub. Bohr was particularly adept at generating unexplained physics. When Rutherford first read Bohr's proposal for how the atom worked, he objected that there was no apparent reason for either limiting electrons to particular orbits or for an electron's choosing one orbit over another. Bohr shrugged off that concern. He did not look for reasons, although he did not insist there were none to be found. In the years since Bohr's theory first appeared, physicists working on the atom had grown used to the fact that there were as yet no reasons behind the basic rules. Bohr's closest disciples, too, were looking for equations that worked, not reasons that explained why. Einstein, of course, wondered why. Questions bubbled persistently on his brain's back burner.

Skreek-skreek-skreek. What fundamentals underlie atomic behavior?

Chatter about the Pirandello play at Max Reinhardt's theater. Why are electron orbits fixed the way they are?

Feel dread over the news that an extremist, nationalist politician has been put in the cabinet. What determines the moment that an atom emits light?

More pressing, however, was the question of whether to explain the Compton effect with light quanta or BKS? Two teams of experimenters were especially important in trying to resolve the issue. In Berlin, Hans Geiger and Walther Bothe, the same pair that had performed Einstein's would-be *experimentum crucis*, went to work testing the rival theories. According to the light quanta theory, light should scatter at the same instant that an electron recoils from the collision. BKS expected that on close examination the two events would not show such a firmly fixed association. In America too, Arthur Compton was still at work. He was preparing an experiment with a colleague to test energy conservation. They planned to use a device called a cloud chamber that let physicists observe free electrons. If energy was conserved, the electrons should move at predictable angles. If Einstein's light quanta idea was wrong, the angles would be random.

Einstein showed no doubts about how these experiments would come out. Max Born even wrote to Bohr in January 1925, months before the results were reported, that Einstein was feeling "triumphant." Others were more troubled. Wolfgang Pauli wrote a friend, "Physics is very muddled again at the moment; it is much too hard for me anyway, and I wish I were a movie comedian or something like that and had never heard anything about physics." In the Netherlands, Ehrenfest was moderately optimistic. He loved Einstein and was utterly charmed by Bohr, so he tended to be neutral in their disputes. Yet he wrote Einstein, "If Bothe and Geiger find . . . a *correlation* [between electron and the scattered light quantum] it will be a triumph of Einstein over Bohr. This time, as an exception [to the usual neutrality] I firmly believe you are right, and I would therefore be happy if the correlation were to be demonstrated."

After 20 years of Einstein's insistence that his light quanta hypothesis was sound, it was time to look at everyone's cards. The bets were in. There could be no more posing, no more bluffing. Ehrenfest's support, Einstein's confidence, Pauli's nervousness could contribute nothing to the outcome. The gambler's emotion while waiting for the ball to drop into a roulette number was caught by Count Kessler. Over tea that February he told Einstein that he seemed to be an unusually successful scientist.

Einstein replied that he was merely unusually lucky. He knew many thinkers who were just as bright and suited to make major discoveries, but were less fortunate.

Kessler held his ground, saying that Einstein seemed "to have some special sort of feelers to tell him where the solution to a problem lies."

Einstein usually did not stress luck's role in science, but Bohr, too, was widely respected for his antennae that smelled out a solution. Bohr had solved perhaps the only problem that Einstein had ever rejected because it looked too hard. Late in his miracle year of 1905, when everything he examined turned to gold, he wrote one of his friends, "There is not always a subtle theme to meditate upon. At least not an exciting one. There is, of course, the theme of spectral lines, but I do not think a simple connection of these phenomena with those already explored exists; so that for the moment the thing does not show much promise." Eight years later Bohr explained spectral lines by linking quanta and the Rutherford atom; so Einstein knew that special feelers alone were not likely to make him triumphant over Bohr.

For months Berlin was rife with rumors that the Bothe-Geiger experiments supported Einstein, but the results were too important to rush. They had to be studied carefully, checked for accuracy, and checked again. It would be intolerable to declare one side victorious and then have the other side win, so to speak, on appeal. Early in April 1925, however, Geiger sent Bohr a note, warning him that his experiment did not support BKS. The scattered light and bouncing electron acted simultaneously, permitting a cause-and-effect explanation for the event and being much too perfectly coordinated to result by chance.

Bohr replied promptly, thanking Geiger "for the great friendliness with which you have informed me of your important results." He added, "I was completely prepared [for the news] that our proposed point of view should turn out to be incorrect. The whole thing was more of an expression of an attempt to achieve as great as possible application of classical concepts, rather than a completed theory." It was good of Bohr to state his motive for the record: to save as much of classical theory as he could. Then he seemed to balk, ending his note to Geiger by saying, "In general I believe that [many] difficulties so far exclude the maintaining of the ordinary space-time description of phenomena, so that in spite of the existence of coupling [between scattering and electron motion], conclusions based on an eventual corpuscular nature of radiation lack a satisfactory basis."

Bohr always was a wordy fellow, but he seemed to be saying that despite the experimental results he still could not accept Einstein's light quanta. Nine days later, however, Bohr appeared to have decided against guerilla resistance. He added a P.S. to a letter sent to a friend at Cambridge, "Just this moment I have received a letter from Geiger. . . ."

How dramatic; too bad for our story that we know Bohr had responded to Geiger more than a week earlier.

Bohr continued, ". . . It seems therefore that there is nothing else to do than give our revolutionary efforts as honourable a funeral as possible." In the end, Compton's data and Geiger's results—not to mention the accuracy of Einstein's photoelectric law—were just too many facts to deny. Reject that authority and his fellow scientists might have stripped Bohr of his union card.

So BKS was dead. Almost exactly 20 years after publishing his

light-quanta paper, Einstein's revolutionary idea had become as orthodox as republicanism. Science was just going to have to make room for
wavy little chunks of hv. Compton's cloud-chamber results came only
that summer, so they were anticlimactic, but they too supported
Einstein by showing that the angle between scattered X-ray and re-
coiling electron matched exactly the requirements for conserving energy and momentum.

Pauli, who had been in such despair that he wanted to join Charlie
Chaplin, was deeply relieved. Rather tactlessly he wrote to Kramers, "I
think it was a magnificent stroke of luck that the theory of Bohr,
Kramers and Slater was so rapidly refuted by the beautiful experiments of Geiger and Bothe, as well as the recently published ones of
Compton. . . . Many excellent physicists would have maintained this
theory and this ill-fated work . . . would perhaps for a long time have
become an obstacle to progress in theoretical physics! because it moves
in a completely false direction. . . . It can now be taken for granted by
every unprejudiced physicist that light quanta are as much (or as little)
physically real as electrons."

Bohr can hardly have been happy when Kramers reported what
Pauli had said, but Einstein was delighted and gleefully told Ehrenfest,
"We both had no doubts about it." Yet Einstein was not content. He
still could not conceive how to unite a wave with a particle, nor, apart
from Compton scattering, could he yet predict when or in what direction a light quantum was apt to move. Other physicists were more
sanguine about these mysteries. A promising young American named
John van Vleck remarked, "It is not surprising that paradoxical theories
are required to explain paradoxical phenomena," but in Einstein's eye
all phenomena were paradoxical until a coherent theory explained
them.

And Bohr had not completely surrendered. From mid-1925 on-
ward he accepted light quanta, but his main feeling was that BKS's
radical rejection of cause and effect had not been revolutionary
enough. He had told Geiger that with BKS he had tried to apply
"classical concepts"—meaning distinct waves and distinct particles
moving through measurable space—as much as possible. Now he
threw that conservative ambition out the window. In a paper pub-

lished that summer Bohr added a postscript saying that after BKS's failure "the generalization of classical electrodynamic theory . . . will require a sweeping revolution." *Electrodynamic theory* was jargon meaning that, with the success of light quanta, Bohr now expected a full revision of ideas about space, time, and relativity.

Part II

A Radical Theory Created

15

Something Deeply Hidden

öteborg, in Sweden, waltzes with surprises. Expecting ski trails
or high-walled fjords, strangers find meandering water instead. A river lazes its way across a Scandinavian steppe flat
enough to be in Holland or somespace east of the Urals where
Mongols thunder through. The slow moving river opens at Göteborg
into a bay, Sweden's busiest port. The terrain that Einstein found in
July 1923 when he arrived to give his delayed Nobel lecture was as flat
as Berlin. During the previous two years, while Germans were nursing
hatreds and ruination, Sweden had been happily preparing for a tercentenary jubilee exposition to celebrate Göteborg's founding in 1621.
New plazas, streets, and buildings were erected to house a festival that
lasted from May through September.

Einstein arrived at the height of the celebration to speak before an
audience of 2,000 souls, including Sweden's King Gustav V, in the new
Memorial Hall. The event was an odd combination of the lost and the
new. The Nobel Prize ceremony was a vestige of the lost civilization
where borders mattered less than imagination. Meanwhile, the
Göteborg jubilee honored the new—the secure frontier and the boasting about what had been accomplished behind it.

The city's buildings were low and ordinary with nothing like the
grand monuments that had been scattered across Berlin during the

half century before the Great War. Göteborg's soil was to blame. It crunches like strips of dried pasta and cannot support tall structures. The most impressive buildings to greet Einstein had been constructed for the exposition. A new central plaza adorned the festival grounds and led to the exposition's buildings. The plaza boasted a fountain with Poseidon wearing a conch cap above a Scandinavian face. Water splashed merrily about him as he clutched a shark in one hand and held a basin high with the other.

Einstein arrived in his accustomed role of reason's champion, but the text he carried in his pocket was unusually dense. Typically, the difficulty of his lectures came from the unfamiliarity of his mathematics, or the listener's unwillingness to surrender some complicating idea, or perhaps the novelty of Einstein's concerns, but this time his density sounded like a parody of Bohr's style. He used five words where one would do, added mathematical jargon he could easily have avoided, and he provided no signposts pointing toward where he was headed or how his thoughts moved. Even the lecture's climactic finish was filled with obscure mathematical references that hid the talk's surprisingly troubled emotions. The crowd that cheered Einstein as he strode past the Poseidon fountain probably did not know how unfairly challenging the lecture would be. Nonetheless, the handful among the audience able to follow the presentation would hear an account of what Einstein looked for in physics and where he wanted to go.

Surrounding the plaza, besides the hall where Einstein headed, rose two other exposition centers, Göteborg's new art museum and a huge wooden building called Machinery Hall. The latter's gothic roof was said to be the largest vaulted roof ever constructed. Its vast display floor overflowed with products ranging from the very old to the most up-to-date. It showed, for example, a history of cutting tools that ran from a stone age "ax" to a modern surgical scalpel made of the finest Swedish steel. Einstein's hosts led him straight into Memorial Hall, past the many portraits of Göteborg's leading citizens, and into the auditorium where he was richly applauded as he came into view.

The Nobel Prize statement had specifically mentioned the success of the photoelectric law and during the eight months since the prize's announcement, Compton's bullet had struck home. Einstein might, therefore, have seized the opportunity to rebut Bohr's own Nobel lecture and assert the validity of the light quanta. Instead, he said nothing at all about the photoelectric effect. He simply laid out his own agenda along with a description of how he could satisfy his ambitions.

His chief tasks were same ones that had consumed him when the World War ended: understand quantum radiation and unify his theory of relativity with the facts of electromagnetism. In the final words of his lecture he reported that he was ready to abandon many ideas to grasp the "most profound physical problem of the present time," quantum theory. Relativity, he admitted, had so far been "ineffectual" at providing insights into the quantum's nature and he could even imagine that some day the "solution of the quantum problem" would lead to "a complete change" in the understanding of space and time. If that happened, his great laws of relativity would be reduced to "limiting" equations that were valuable in solving many practical problems but do not describe nature's foundations.

With his closing, Einstein had arrived at a radical level that Bohr would not reach for two more years. Only in 1925, after BKS's failure forced acceptance of the Compton effect and light quanta, would Bohr concede that this result meant revolution. Einstein had sensed the revolution's coming for years, and any uncertainty was settled after the Compton discovery. Light quanta—not just the mathematical abstraction $h\nu$—but the light quanta themselves had become so important that they were soon given a sleeker, less clumsy name—photons—and as photons they will be known through the rest of our history. This new fact of nature that Einstein had called "revolutionary" back in 1905 would finally trigger the quantum revolution that was to begin in 1925.

Photons had already brought a technical revolution, changing the lives of millions of unscientific people, as could be seen plainly at Göteborg's festival of progress. Electric lights illuminated the festival's auditorium with trillions of photons. Introduced by Edison only 44 years earlier, they used a current that somehow stimulated a hair-thin filament to emit a steady blaze of light. Meanwhile, hospitals were using X-ray photons to study patients and their broken bones. Broadcasting, too, had come of age with towers emitting photons to send wireless messages, permit ship-to-shore communications, and entertain and inform the public through radio programs. The age of the photon tool had arrived and anybody visiting Göteborg's Machinery Hall could see that once something becomes a tool—be it made from stone, bronze, or steel—innovators find more and more uses for it. So the coming era of television, lasers, X-ray telescopes, compact disks, CAT scans, and a thousand other photon-based devices was implicitly promised in Göteborg on the day Einstein spoke there.

With the photon already established as a fact of technology, of nature, and of daily living, the question Einstein addressed in his lecture—*What now?*—was one faced in one way or another by many members of his audience. For the practical people who were inspired by the displays in Machinery Hall, new facts of nature call for new inventions. For them, understanding was never the point behind learning; *using* was the point. You could see that in the story of the compass. Vikings had transformed a mystery into an invaluable arrow that pointed their way across the featureless ocean. The compass served for centuries without its users ever gaining a clue as to how it worked. Even in 1923 magnetism was still not deeply understood.

Novelties can make the artistically inclined want to capture exactly how a new wonder changes the look of things. Each of the paintings in the Memorial Hall's portrait gallery was the residue of that imaginative wonder and reflected an hour when an artist peered as attentively as possible at the facts of a single face. No less than experimentalists, artists wanted to know exactly what it was before them. "Nature for us men is more depth than surface," Paul Cézanne had written, and Einstein would have agreed completely, but then the artist went on to advise introducing "into our light vibrations, represented by reds and yellows, a sufficient amount of blue to give the impression of air." For artists the point of getting the facts right is to express what they see there. For them a compass was a detail to be seen exactly and portrayed precisely. Is there a clear piece of glass above the needle? And does it betray its presence by slightly altering the look of what's behind it? Good, then get that subtle difference on the canvas.

Einstein's reaction, a confident effort to understand nature's power, defined scientific wonder. For Einstein, something like a compass was a clue, a witness to be brought before the bar and cross-questioned as to its meaning. Einstein's father had in fact shown him a compass when he was a small boy: "I can still remember—at least I believe I can remember—that this experience made a deep and lasting impression on me. Something deeply hidden had to be behind things." At age five he might have not yet felt that anybody could discover the deep mystery behind this wonder, but somehow in the years since he saw his first compass Einstein had grown as confident as Sherlock Holmes that he could follow clues to their solution. He was no optimist about human affairs, but when it came to his own abilities he looked only for the heaviest lifting. The wondrous fact of the compass was that it moved without being touched. There must be something hidden but real, lying invisibly in empty space, something ready to spring into action whenever a compass goes by. He could have reverenced that mystery, or enfolded it into an art, or made practical use of it, but because he was a scientist by nature, Einstein sought simply to understand it, and in his Nobel lecture he told his audience exactly what was involved in that understanding.

It might have been entertaining to ask King Gustav what Einstein had said. Perhaps he would have been as quick-witted as Count Kessler and remarked that he caught the significance more than the meaning. Too bad, for the lecture was unusually detailed about how Einstein approached a mystery and struggled to take the step necessary to approach understanding.

Once Einstein had entered the auditorium and been applauded so roundly that the audience finally lost its need to clap further, the celebrated hero began to lay out what underlies "bona fide scientific knowledge."

Facts of nature, he told the audience determine what concepts and distinctions will be allowed into a theory. Facts are a theory's gatekeeper. It is not that scientists look harder than artists, or explore more fully than artisans and engineers. The secret behind Einstein's scientific imagining was that he was always looking for meaning, which he defined as "the extent to which observable facts can be assigned to" concepts and other abstract forms "without ambiguity." Facts without concepts, Einstein once complained, produced "a catalogue and not a system." On the other hand, a system without facts was not science. From his earliest days as an unemployed graduate in Switzerland, Einstein insisted that physics rested on concrete facts and that mathematical conceptualizing was only a means for expressing the laws governing these facts.

Logically, Einstein conceded, the problem of understanding was badly tangled. First, facts authorize concepts and give them meaning, but then concepts justify and explain the facts. A scientific theory becomes a self-supporting arch in which the factual pillar supports the concepts and the concept pillar supports the facts. The "step" that Einstein kept seeking was the discovery of an arch's keystone that would allow facts and concepts to stand together by themselves.

Einstein saw the logical risks of this approach. It lacked the grace of Euclid's enviable geometry, which pulled an entire system of knowledge out from a few self-evident truths, and he apologized for this gracelessness to his listeners, commenting, "We are . . . not sufficiently advanced in our knowledge of Nature's elementary laws to adopt [a logically] more perfect method without going out of our depth." Did King Gustav mutter to himself *I'm glad he cleared that up?*

The imaginative steps that Einstein took always managed to snap both pillars of a logical arch together. Snap, the fact tells us *this* about the concept and, pretty much simultaneously, snap, the concept tells us *that* about the fact. Compton's discovery had shown the process. It is a fact that X-rays change frequency when they lose energy; Einstein's concept of light particles says a photon's energy changes as its frequency changes. So, snap, Einstein's concept explains the change in X-ray frequency, while, snap, the fact of a change in frequency justifies Einstein's concept. This snap-snap process lay behind much of Einstein's success. He would struggle for years thinking about some paradoxical fact of nature and then, snap, realize that the fact implied something quite unexpected about a concept previously taken for granted and, snap, he would recognize too that the revised concept implied many things about the fact. Suddenly, in a tsunami of understanding, the years of puzzlement would be followed by a few weeks of intense labor that transformed the paradox into a coherent theory.

In his Nobel lecture, Einstein described the union of facts and concepts that led in 1905 to his creation of the special theory of relativity and then, 10 years later, to his generalizing of the theory. He concluded by telling his audience something of his interests in taking general relativity another step and uniting it with electromagnetism.

"The mind striving after the unification of the theory . . . ," he told the audience, generalizing again from himself. The whole phrase could have been replaced by the lone word "I." If he had said that, more of the audience might have realized that he had shifted into a personal account of what interested him in science. Einstein told the room that he "cannot be satisfied that two fields should exist"—the gravitational field and the electromagnetic one—"which by their nature are quite independent." He was seeking "a mathematically unified field theory . . . in which the gravitational . . . and the electromagnetic field are interpreted only as different" elements of the same thing.

This unification would be like the unity Einstein had already found when he showed that matter and energy are different forms of the same thing. It was also the kind of solution he wanted in the problem of photons. The Compton effect had shown that light is just as much a particle as it is a wave. He needed to adjust his concept of a wave so that the facts of light's nature would become coherent. With that step, light's wave-particle duality would change from being a paradoxical fact to being an intelligible one.

A problem for the unification of gravity and electromagnetism was, as Einstein told his audience in Göteborg, that gravity and electromagnetism do not contradict one another. They ignore one another. Once they had been contradictory, but with relativity Einstein himself had exorcised those paradoxes from physics. Now gravity and electromagnetism sat together like two strangers on a bus seat, paired and yet having nothing to do with each other. Einstein's gravitational equations worked fine; so did the established equations defining an electromagnetic field. With neither contradictions nor harmonies to suggest a point where the two might arch together, Einstein told the people of Göteborg, "We are restricted to the criterion of mathematical simplicity."

Did anyone among the 2,000 listeners come to attention here? Possibly not, but it is more agreeable to suppose that somewhere among all those heads was a brain alert enough to ask if Einstein was bidding farewell to the scientific quest. He had been saying throughout the lecture that scientific knowledge demands meaningful concepts. Mechanics before relativity, he said, had not always been meaningful. The mathematics had seemed clear enough, but on close examination it turned out to rest on rules and assumptions rather than facts of nature. When it came to postwar theories of gravity and electromagnetic fields, however, the philosophizing had been squeezed out of the equations. Everything rested on facts. But, Einstein said, there were no facts linking nature's controlling fields. Simplicity as a goal, Einstein warned, "is not free from arbitrariness."

There now, that one alert head could relax a bit after all.

Einstein was not abandoning science by trying to get away from the facts. He realized full well that he could not introduce any new concepts into his work. There could be no redefining time or proposing novelties like photons. He was still trying to explain nature in nature's own terms. Einstein had returned to his familiar post as lone observer high in a crow's nest looking at storm clouds while everybody on deck sees only open skies. As usual his cry of "Storm ho" was greeted by the officers on deck with skepticism and even a bit of gleeful contempt. There had been something satisfying in laughing at the great physicist's talk of photons, but then the Compton effect wiped away the grins. Now Einstein was back on his lone perch, insisting there was a problem, and none took it seriously.

Of course, few, if any, members of the Göteborg audience realized how much thought lay beneath Einstein's dense jungle of prose. They cheered anyway, possibly even applauding all the more loudly the less they understood. The point of sitting there was to be present when the prophet spoke. Nobody understood the Oracle of Delphi either, but that did not mean that nobody came to listen. So they applauded and no doubt decades later were still telling people they had heard the great Einstein give his Nobel lecture.

—*Oh, and did you understand it?*

—*Not one word,* laughs the boaster, still proud of having heard it and quite unembarrassed about having understood none of it.

It had not always been like that. A generation earlier physicists like Mach, Helmholtz, and Maxwell had given great popular lectures to ordinary working people who had understood the talks and been interested. Helmholtz, missionizing for science, preached, "There is a kind, I might almost say, of artistic satisfaction, when we are able to survey the enormous wealth of Nature as a regularly-ordered whole— a kosmos, an image of the logical thought in our own mind." And Mach assured his audiences that a popular expression of the time— *ignorabimus*, we can never know—was a falsehood and that everything about the world could be learned "entirely by accurate observation and searching thought."

Something had happened since those days of public optimism. Maybe the science had gotten harder, but mostly it was the great folly of the war and the increasing professionalization of science. It is telling that Thomas Mann chose that year (1923) to write that although he knew and understood "very little about the famous Mr. Einstein," he had the clear impression that the boundary between physics and metaphysics had become more tenuous. Actually, the opposite was happening. Einstein had chased metaphysics out of the story of motion and gravity, but he had retained and reinforced the Helmholtz view of "Nature as a regularly-ordered whole." Yet, the general public no longer expected to be able to follow the explanations and did not realize that Einstein was clarifying, not muddying. People were good humored about it, laughing at their own incomprehension, and thanking heaven that there was a hero like Einstein who did understand it. But they missed Einstein's larger sermon that no matter what your vantage point, the cosmos follows the same natural laws.

The hero made his way out of the auditorium and moved back past the portraits of distinguished Göteborgians. Sunlight lasted 18 hours that day, providing ample time to do any sightseeing and socializing. Eventually he boarded a train for the south and followed the coast toward Malmö. On a ferry he put Sweden behind him as he steamed toward Denmark. It was not long before Copenhagen appeared as a smoky cloud, and then as factory chimneys poking above the horizon. Beneath the smoking chimneys as Copenhagen's greencopper domes and red roofs rose into view, Niels Bohr was waiting for Einstein to land.

16

Completely Solved

ohr and Einstein were already deep in conversation when they
boarded one of Copenhagen's trams. Although Bohr enjoyed a
good quarrel, Einstein preferred to pass by fools in silence and
rarely rebutted even published opinions. "I do not have to read the
thing," Einstein once said of a book that denigrated his role in creating
relativity theory. "If he manages to convince others, that is their own
affair." So it was a mark of great admiration that Einstein was willing
to talk endlessly with a man he disagreed with so deeply. Bohr was
always looking for converts; Einstein was not.

What Einstein wanted was encouragement and new ways of
thinking. "My friend and colleague M. Besso," Einstein had written at
the end of his first relativity paper, "steadfastly stood by me in my
work on the problem discussed here, and I am indebted to him for
several valuable suggestions." *Steadfastly stood by* . . . that was the kind of
thing Einstein appreciated, and when you look at the major people in
Einstein's life—Ehrenfest, von Laue, Lorentz, Planck, and Elsa, too—
the quality they all shared was a steadfast support for Herr Professor
Einstein.

Bohr was not that way, but Einstein appreciated the vigorous
workout given his ideas whenever the two men talked. We naturally
focus on imagination's grand moments, the steps and insights that cre-

ate new things, but the long talking and wrestling was critical. Einstein told a friend it would be wrong to consider his great step in relativity "as a birthday, because earlier the arguments and building blocks were being prepared over a period of years, although without bringing about the fundamental decision." For Einstein, the long tram talk with Bohr was not idle. He was always preparing his garden, always looking for the insight that would make sense of two contradictory ideas.

Bohr, too, disliked contradictions, although he was closer in spirit to Ernst Mach's worries than to Einstein's. Mach used to complain about the contradictions between the ways people talked about physics and about themselves. Bohr, too, wanted a unified way of talking about things, which is not quite the same as saying the things themselves are unified.

The contradiction that Einstein had once faced when he worked on relativity was as baffling as those that later plagued the quantum. The laws of motion were well established, but they contradicted demonstrated facts of electromagnetism. He had titled his relativity paper, "On the Electrodynamics of Moving Bodies," indicating that it would be about the way electrical bodies moved. In 1905, when Einstein published that paper, the electron's discovery was less than 10 years old, and questions about its movement were in the avant-garde of theoretical physics. The leading thinker on the subject in those days was Hendrik Lorentz, a Dutchman already past 50 and yet still the trail scout who found the clearest routes through the wilderness. In old age, Einstein often said relativity would have been impossible for him without Lorentz. Indeed, into the 1920s many physicists still referred to relativity as "Lorentz-Einstein theory."

Bohr's work also focused on how the electron moved. Bohr's approach, however, saw quantum jumps as intrinsic to the story while relativity rejected the possibility of discontinuous changes that Bohr demanded. Even so, many physicists hoped to see the theories united. In 1916, Arnold Sommerfeld had shown that by including relativistic mathematics in Bohr's theory he could solve certain puzzles that Bohr's equation alone could not calculate.

Einstein never seemed much interested in Bohr's side of electron movement. Although Einstein's explanations could be very abstract, the phenomena he started with tended to be remarkably straightforward. Bohr's work assumed a particular kind of atomic structure and a particular explanation for the source of atomic spectra. Einstein preferred the kind of phenomena that a garage mechanic might notice. For instance, Einstein began his relativity paper with two objects, a magnet and an electrical conductor. They could be found in any commercial garage of that day or this and yet, when used together, they produce contradictory results. When motion produces an effect, physicists do not care what moves and what stays still. If you strike a match against a matchbox, you get a flame. Also, if you hold the match steady and rub the box against it, you still get a flame. If wind moves through a wind chime, you get a sound. Likewise, sound comes if the air is still and the wind chime sails through it. We would be startled if wind caused the chime to sound, but moving the chime caused it to glow. Yet conductors and magnets produce exactly this kind of surprise. If a magnet moves past a resting conductor, you get one effect. You get a different effect if the conductor moves and the magnet lies still.

The first effect is an electric field measured as energy. This energy field appears around a magnet if you move the magnet near a resting conductor.

The second effect, known as electromagnetic induction, produces an electromotive force, measured in volts. When a conductor moves past a resting magnet, this electromotive force (voltage) appears in the conductor and begins "pushing" electrons.

Both of these effects were well known before Einstein's day. They were discovered and defined by two heroes of nineteenth century British physics, Michael Faraday and James Clerk Maxwell. Einstein's father had been in the electrical equipment business, producing dynamos and other machines, so this matter of electrical fields and forces was old news in the Einstein home. But Einstein, unlike the majority of physicists, worried about this "asymmetry" in which one motion produced one effect while an equal and opposite motion produced a second effect. Turn-of-the-century physics could not account for such a contradiction.

Several times before in the history of science the theory of motion had needed enlarging. The ancient Greeks produced a notorious error remembered as Zeno's paradox that purported to show motion was impossible. To get from my bed to the doorway, the argument went, I first have to cover half the distance between the two points. Then half the distance again. Then half again . . . and so on forever. A person can never reach the goal because there is always another half distance to cross. It requires a certain temperamental attachment to reasoning before you can take this argument seriously, and most people probably dismiss it out of hand. Clearly, we do not really move in that half-again style. But the challenge is in coming up with some better account of motion. Where had Zeno gone wrong?

Zeno was philosophically attracted to the idea that the world is stable and all change is illusion. He used his paradox to "prove" that although the world seemed full of change and motion, it really stayed the same, and he felt content to keep the argument as it stood. It was a profoundly unscientific argument because it sought to deny the facts of experience. It did not look for a deeper truth that explained the facts, but instead required a faith in logic so strong that a person would accept it instead of everyday experience. A physicist like Einstein would never be content to simply defy the facts, but as a believer in logic, Einstein also would have refused to brush Zeno's reasoning aside.

To overcome Zeno you will get nowhere by looking for a flaw in the logic he used. In these days of moving film we know full well that motion can be broken down into an infinite number of steps. There is no theoretical limit to the number of movie frames we could use in capturing the motion from bed to doorway. Instead of looking for a logical error, the solution was to find a deeper idea that Zeno had left out. At first glance, movement may seem to be merely a change in position. A second look, however, shows that Zeno overlooks part of the story. Motion involves both a change in position and a change in time. Zeno's argument, Einstein would have said, was not illogical, it was incomplete.

True, I could go halfway between bed and doorway and stop so that no further change in location occurred, but I cannot stop time's winged chariot. Take another look at that movie reel. Each frame shows a snippet of motion through space, but any two frames show a span of time. Zeno might have cried, wait, *time is an illusion too.* Before you can have a second, you need half a second and then before you can have the last half second you need a quarter second . . . and so boringly on. But now we have caught Zeno in his error. He insists on separation instead of unity, on arguing only about space or only about time when we know that motion concerns both space and time changing together. No matter how short a space we consider, there is always a speck of time changing as well, so that even if we are down to tracing a millionth of an inch, we still have the change in a millionth of a second (or whatever) giving us a rate of change. If the rate is kept up, the traveler will arrive at its destination. Zeno might have overlooked the point, but Archimedes knew it well enough. Movement is measured by distance *and* time. The phrase "60 miles per hour" brings space and time into a single concept. The point had been settled before the Christian era began.

The Copernican theory of a moving earth forced some still-deeper thinking about movement. If the earth is really spinning at a speed measured at the equator of about 1,000 miles per hour, my ability to drop a ball on my foot becomes a puzzle. Let's say I hold a ball about 5 feet above my toe and drop it. It takes perhaps a quarter of a second for it to land. During that time the earth rotates about 367 feet, yet the ball still hits my toe. Why doesn't it miss? Many people today secretly wonder about this sort of thing. When I ride in an airplane going hundreds of miles per hour, why doesn't the back of the plane swat anything thrown in the air?

Galileo solved this one by adding what we can call relative systems to our idea of motion. Einstein called them coordinate systems and many physicists call them inertial systems. The name does not matter so much, but the point is crucial. Motion can appear differently to different observers, depending on whether they share the same relative system. Galileo showed what he meant by using a ship as an example; we can do the same, modernizing the armada.

Suppose some sailors play baseball on the deck of an aircraft carrier. The pitcher throws the ball. Behind the catcher is a baseball scout with a radar gun that measures the ball's speed at 80 miles per hour. It happens, however, that the carrier is passing a lighthouse where another fellow with a radar gun measures the same pitch and gets 100 miles per hour. The extra 20 miles per hour was added by the carrier, which is cruising at that speed.

The players and scout aboard the carrier are together in the same relative system. They are all at rest relative to one another. For them, things move as though the ship really were at rest. The same principle applies to a rotating earth and flying airplanes. I don't miss my toe because both my hand and foot share the same relative system. I don't have to take my foot's motion into account when I drop the ball because, as far as I am concerned, my foot is not moving. Likewise, the back of the plane and my seat share a relative system. If I toss a magazine to a person across the aisle, I need not worry about the way the back of the plane is moving because, from my perspective, it is not moving.

Getting back to shipboard baseball: the fellow in the lighthouse does not share a relative system with the aircraft carrier, so he measures the effect of the ship's motion on the ball. Likewise, an observer in a spacecraft watching me drop a ball onto my toe would see the ball arc 5 feet to the ground while flying more than 100 yards eastward along with the spinning earth.

Linking time to motion has an intuitive feel to it, making Zeno seem a little silly for having forgotten it, but Galileo's relative motion does not harmonize so readily with our psychology. We don't see ourselves as perpetually at rest. Even Einstein and Bohr, talking eagerly, paying no attention to the city outside the window, perceived (if only they had looked) the tram as moving through Copenhagen and not vice-versa. Nevertheless, physics in Galileo's day divorced psychology and made it an established principle that uniform motion is always relative. Something is moving, but whether it is one thing or the other is just an arbitrary choice. But Einstein had noticed a simple motion involving magnets and conductors that seemed absolute. You get different accounts of what happens depending on what you view as being at rest.

Einstein's instinct was to deny that this contradiction could be correct. He needed a rule that overcame electromagnetic logic, so he asserted what he ever after referred to as the "relativity principle." According to this notion, "the same laws of electrodynamics and optics will be valid" in all relative systems where Newton's laws of motion are valid. The principle was not a theory or hypothesis so much as it was a prayer. If science was to work, its laws have to be the same in every relative system. Otherwise behavior will appear random. Suppose, for example, a scientist conducts an experiment in a relative system where a conductor seems to move past a magnet, but "really" the magnet is moving past the conductor. This scientist will see that when the conductor moves, an electric field appears around the magnet, and the scientist will develop electromagnetic laws that are exactly the opposite from those found by earthling scientists.

On his walks through Bern and evenings chatting with Besso, Einstein worried for years about such differences and the problem of making Newton's mechanical rules fit together with the laws of electromagnetism. The persistence of his wonder suggests that either he was a young man of astonishing faith in nature's lawfulness, or he was a crazy man unable to stop charging windmills of his own invention. Possibly right from the beginning, Einstein had guessed the solution—electromotive forces are really just electromagnetic fields with a 0 energy measure—but like Detective Columbo, who always guessed the culprit at the beginning of his investigation, he still had to prove it.

In Copenhagen, crossing the tracks, Bohr and Einstein waited for a return ride. Fifty years earlier it would have seemed impossible that the world's leading physicists would still be disputing the nature of light. The matter had appeared settled in 1873 when Maxwell joined electricity, magnetism, and light into one unified phenomenon—as decisive an achievement as uniting distance and time. It had seemed then that centuries of dispute about light had been resolved.

Galileo had raised the question of whether the arrival of a flash of light is "instantaneous or whether, although extremely rapid, it still occupies time." He even proposed experiments to measure the speed of light. Not until the nineteenth century, however, could machines be made precise enough to allow the experiments. The research had established that light travel does take time, but the same work led to some other questions. In particular physicists wondered why light's relative speed was not as easy to detect as relative motions on a ship deck. A Frenchman named Armand Fizeau measured the speed of light in water and found that the expected equations for relative speed were not as straightforward as those for a ball thrown on a ship's deck. Two Americans, Albert Michelson and Edward Morley, conducted very precise measurements that showed no change in speed relative to the ether that carried the light wave. Experiments like these were complex and subject to interpretation, but the results were plainly not the simple ones expected.

Einstein, when he was 16, had also recognized something odd. He had imagined what it would be like to travel at the speed of light while chasing after a light beam. When two streetcars travel side by side at the same speed, they become part of the same relative system and don't appear to move at all. The passengers in separate cars can talk with each other or even play catch if they have a mind to, exactly as though the trams were resting side by side in a railway yard. In his mind's eye, Einstein saw what a motionless beam of light would look like when viewed that way, but he realized that no such thing as standing light ever turns up in nature. There were no experiments where light could be brought to a standstill. Furthermore, Maxwell's equations would not work if the speed of light were equal to zero.

That imaginary ride appears to have been the first time Einstein began chasing after a serious split between logic and nature's laws. It marked the birth of his lifelong technique; find the point where two sets of ideas rub against one another and then discover what the existing theory leaves out. When Einstein and Besso talked together, Einstein was looking for a way to unite electromagnetic and mechanical motion, just as, 20 years later, riding back and forth through Copenhagen with Niels Bohr, he was looking for a way to unite waves and particles, or gravity and electromagnetism.

Bohr, for his part, was still resisting the photon, despite the Compton effect, because it was going to force disunity on physics. Light would not be one thing *or* its opposite; it would become one thing *and* its opposite. Mach had worried that people talked one way when they discussed physics and another when talking about anything else. Now Bohr feared people would be forced to talk about physics one way, and then, without even changing the subject, talk in another way. The solution, Einstein felt, was to find out what light really is. Bohr thought the solution was to find a coherent way of talking about light.

Most other physicists were not so single-minded about how an explanation should work. Their approach was professional, grabbing what worked and letting the philosophical side of the question look after itself. The generation of physicists before Einstein did well with that professional approach. They had to resolve the mystery of the electromagnetic fields that are inherent in Maxwell's equations. Most physicists had a hard time grasping the idea of a physical field, a region of empty space that somehow produces effects. It was Lorentz who clarified the concept by stressing two realities—electrons and the ether. Electrons were theoretical atoms of electricity that carried an electric charge. They existed in Lorentz's theory years before the English physicist, J.J. Thomson showed them to be fact. Equally important to Lorentz was the ether that supported light. The ether had no normal physical properties and was difficult to detect, but it permeated everything. "Empty space" really did not exist. Every point contained the ether. Furthermore, the ether provided a way to see beyond relative motions to absolute movement. Ships move one way, planets go another, but the ether is not moving anywhere. Motion relative to the ether is true motion because it is relative to the whole universe.

For Lorentz, therefore, the Michelson and Morley experiments that failed to find an ether were a serious problem. He saved his theory by proposing that movements against the ether force objects to contract. The molecules literally press together. Normally these contractions would go unobserved because everything, including measuring sticks, contract at the same rate. They became apparent only when measuring the speed of light because the contractions eliminated the relativistic velocities of light, creating the illusion that light moves at the same speed regardless of relative systems. Lorentz also produced a set of equations that compared the sizes of normal and contracted objects, publishing them in final form in 1904.

This work by Lorentz led some to deny that Einstein discovered relativity. Iconoclasts, anti-Semites, and congenital contrarians are al-ways "discovering" Lorentz's theory. Many years later, Serbian nation-alists would even argue that Einstein's first wife was the real genius behind relativity and that Einstein had simply appropriated her labor. Contemporary historians and biographers know they can get into trouble by dismissing charges against their main figures. The scholars who devoted their careers to studying Thomas Jefferson look foolish for having rejected out of hand the claims that their man used one or more of his slaves sexually. An example like that can frighten other writers into hesitant "balance," making wishy-washy stands: *Well, of course there is no clear proof in Mileva's favor, but neither is there a smoking gun against her.* You cannot go wrong that way. Except there is a smok-ing gun. It is the style of Einstein's thought.

Einstein was, his whole life through, obsessed with ironing the logical contradictions out of physics, especially the contradictions re-lated to electromagnetics and particle mechanics. Mileva had no such obsession. The issues Einstein tackled in 1905 recurred in 1907 and in 1909 and in 1915, 1916, 1923, and on and on. They were part of his fabric. By contrast, if Einstein had, in 1904, somehow produced the Lorentz paper, we would know something was wrong. Even though Lorentz's math is very similar to Einstein's work in 1905, its reasoning was most unEinsteinian. Yes, the Lorentz theory was visualizable and Einstein liked visualizability. Lorentz's ether forced molecules closer just the way a hand squeezes together the holes in a sponge. The contractions were fully predictable in a mathematically precise way. They were also absolute. Lorentz's equations predicted contractions along all three dimensions as they would appear from the ether's ob-jective viewpoint. One thing, however, became unknowable, unvisualizable. That was the ether itself. Its effects were clear, its im-portance undeniable, but movement through the ether became in prin-ciple beyond all measuring. Movement relative to the ether forced a contraction in the very devices needed to discover these movements. The ether became like one of God's angels, acting everywhere while discoverable nowhere. Einstein never liked angels, and if they had ever turned up in one of his theories, we would wonder whose influence we were seeing.

Lorentz's work was typical of a professional scientist. He had a theory. Experiments raised difficulties with a critical part of the theory. He revised his idea to resolve the problem and save the theory, but Lorentz had paid no attention to the issues that haunted Einstein. Lorentz's new equations did not address the contradictions between electromagnetism and ordinary motion, and there was still the business of Einstein's personal vision of the impossible light-beam-at-rest.

Einstein, in thinking endlessly about these puzzles came to a guiding assumption, one that he set right up there with his relativity principle: "light always propagates in empty space with a distinct velocity that is independent of the state of the motion of the emitting body." It is amusing to remember how many street philosophers say that Einstein proved everything is relative when his guiding axioms assert absolutes. The relativity principles asserts that the laws of physics are absolute and do not change as one travels from one relative system to another. The second axiom holds that the speed of light is unchanging whether the light is coming from a speeding ship or a standing lighthouse.

It was a bold idea, but not entirely new. In 1904, Henri Poincaré told an audience in St. Louis that physics needed "an entirely new kind of dynamics which will be characterized above all by the rule that no velocity can exceed the velocity of light."

Although in the spring of 1905, as Einstein chattered with Besso, Lorentz's rules and Poincaré's speed limit had already appeared, Einstein, stuck in Bern, Switzerland, with no decent physics library on hand, knew nothing of those things. Even if he had known he would not have been satisfied. It is not enough to assert that the speed of light is a fixed limit on how fast anything can move; you have to explain how such a rule can make sense. Absolute speed leads to mathematical contradictions. There are good reasons for finding different speeds in different relative systems. If you measure the distance from pitcher to home plate in our aircraft-carrier ballgame, you find 60 feet 6 inches. The speed of the ball equals the distance traveled to reach home plate divided by the 0.52 seconds required for the ball to arrive at home plate. Viewed from a lighthouse, however, the ball travels 75.6 feet—the 60.5 feet to the home plate plus the 15.1 feet that the ship covers while the ball is in flight. It doesn't take too many math skills to see that 60.5 feet divided by 0.52 seconds gives us a different number from 75.6 feet divided by 0.52 seconds. So how can any speed, even a very high one, ever be absolute? Problems like this made Einstein suspect the absolute speed of light was untenable and that he would have to drop it.

Lorentz, of course, had seen the problem and had found a solution, or at least he had found what today's computer programmers call a "workaround." Along with the three equations that showed how objects seemed to change size, Lorentz had a fourth that showed how time seemed to change too. He called these distortions *local time* to distinguish them from true time, but Einstein was never interested in workarounds because they obscure the real meaning of things. Historians who like to play *what if* games can entertain themselves by asking what would have happened if Einstein had been buried in an avalanche during one of his schoolday jaunts up into the Alps. Lorentz had developed a coherent mathematics that produced correct answers for movements involving the speed of light. How long would it have been before somebody bothered to rethink the matter? Would Lorentz's ether have become as embedded into twentieth-century physics as the crystalline spheres had been in medieval astronomy?

As it was, Einstein was still very much alive, still wracking his brains, and still ignorant of Lorentz's 1904 publication. One afternoon he poured out the whole of his troubles to Besso. They went over the paradoxes and contradictions between effects and principles. With it all stretched out before him Einstein at last took his great step. Snap, the relativity principle will hold if time tick-tocks differently for different observers. Snap, the speed of light can be the same in all equations if the pace of time varies between relative systems. Einstein's step was the realization that the pace of time is as relative as distance. If one observer sees light travel 10,000 feet and another sees it travel 12,500 feet, the only way 10,000 feet per unit of time and 12,500 feet per unit of time can equal the same rate is if the times are different for the two observers. Mathematically it may seem obvious that the time part of the fraction must change if you need to calculate the same speed for changing distances, but physically the step took enormous imagination because we have always believed that time is absolute. Just as shipboard and lighthouse surveyors will measure different distances for the same flying ball, so too, timekeepers in different relative systems will measure different times. The distance between events depends on relative systems. The time between events also depends on relative systems.

With the realization that time is as relative as distance, Einstein stepped beyond Lorentz. Lorentz understood what had happened to him and admitted that the source of his "failure" to discover relativity, despite the years of work, had been his "clinging to the idea" of a true time and his thinking that local time was not real. Poincaré had missed it for the same reason, thinking there was one true time throughout the universe. Now Einstein's step was really carrying him toward the invisible face that hid behind the moving compass.

— —

Still in Copenhagen, Bohr and Einstein again made their way across the tracks and again began to wait for another tram. The fact that they were having trouble reaching their destination was no mark against their achievements. From the beginning, a few physicists had recognized in Einstein's relativity paper the work of a new Copernicus. The original Copernicus, too, had defied one of the most universal experiences. The sun can be seen rising and setting every day. Even very bright, very imaginative people responded to Copernicus by demanding to know just how it could be that the sun really stands still and that its motion is a relativistic illusion. Our experience of time is similar. Time flies but keeps to a faithful schedule. It was and is hard to grasp how that steadiness could be just another relativistic illusion. Very few people can abandon their belief in true time. The fact that the watch on my arm runs slow does not persuade me that my time is just as good as anybody else's.

Newton would have said that God knows which time is true.

No, Einstein would have replied, God knows which equations are true. The measurements are transitory and local; the relationships enduring and universal.

So I was right about change being illusory after all, Zeno might have chimed in, allowing Newton and Einstein to join forces and hurl sticks at the ancient Greek.

Einstein was thrilled with his step and the next day said to Besso, "Thank you, I've completely solved the problem." The two concepts—laws of electrodynamics hold true in all relative systems, and the speed of light is absolute—now formed an arch strong enough to support a multitude of shifting facts. Einstein spent the next five weeks in a frenzy of labor, putting the whole theory down on paper. When he was done, he collapsed. He spent the next two weeks in bed while Mileva checked his article closely before telling him, "It's a very beautiful piece of work."

Much of his effort during those five intense weeks had been mathematical, deriving equations that expressed time's relativity. He produced the same four equations that Lorentz had published the year before, but in Einstein's paper their meaning was quite different. Lorentz's equations predicted how objects would contract along their direction of motion; Einstein's told how anything measurable in one relative system would appear to another relative system. Mechanical contraction played no role in Einstein's theory.

Lorentz's theory was dynamic, using the ether to squeeze molecules. Einstein's theory was mathematical, describing what happens but including no angels to make it happen. Lorentz's theory made the universe more complicated; Einstein's made it simpler.

Einstein's simplicity clarified what he meant by reality. Ultimately, the face of truth was natural law, the description of nature's givens and how their relationships make the world appear to its observers. The appearances themselves—say, the speed of a baseball—might vary, depending on local circumstances, but the phenomenon is real. After all, the distance between the baseball and the pitcher does change and no observer will deny that fact, but the appearance's measurable details are relative.

The purely qualitative part of the phenomenon remains a given. If the batter gets a hit, it remains true for all observers. We can imagine a broadcaster moving relative to a pitch in such a way that the ball appears to have been traveling at 125 miles per hour. "Wow!" shrieks the announcer, "How can a batter be so quick?" Meanwhile, another broadcaster in a different observation perch might say, "Well, of course, he hit that ball. The radar gun shows it was only going 35 miles an hour." The measurable parts of the event are relative, facts that change with circumstances, but the collision of bat and ball is one of the givens of existence.

Relative natures are not trivial. If a baseball passes my head at 125 m.p.h., it will sound and look different from one going 35 m.p.h., yet these differences depend entirely on local circumstances. Many things that people have taken to be absolutely real turn out to be such dependent phenomena. The wind we feel when we poke our hand out a moving car's window is another example of a dependent phenomenon. It is so real that even dogs love putting their heads out of car windows to feel the wind bat their faces, but this sensuous wind is obviously created by the moving car. Its properties come from circumstances, not from nature itself. The earth is turning even more rapidly, yet because the atmosphere turns with the earth we do not feel that breeze. The highway's air, however, does not move with the car and seems to fly past us. If the air were moving along with the car, the wind would fade, even disappear.

A car passing a pedestrian at 60 m.p.h. has a special sound, a kind of quick rising thud that fades just as quickly, and there is a brief wind blast as well. When that same car passes another automobile, one doing 55 m.p.h., the sound is different and the wind is different. None of these sounds or winds is an illusion, but we make a serious error about the physics of the case if we treat such dependent phenomena as the independent givens of reality.

The fact that striking effects can depend entirely on their relative systems finally explained Einstein's paradox of magnets and conductors. As he had guessed, the electromotive force that appears when a conductor moves through a magnetic field is a secondary phenomenon. When you solve the electromagnetic equations for different relative systems, you discover that an electromagnetic field around the magnet has no strength when a magnet appears at rest and the conductor appears to move past it. That is not to say voltage is not real. It is as real as the wind your dog loves to enjoy, but it depends on appearances. Every scientist in any lab in the universe will find the same law at work. If a magnet appears to move, the magnetic field appears as well.

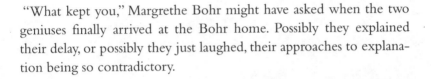

"What kept you," Margrethe Bohr might have asked when the two geniuses finally arrived at the Bohr home. Possibly they explained their delay, or possibly they just laughed, their approaches to explanation being so contradictory.

17

Exciting and Exacting Times

Both Göteborg and Copenhagen lay at Einstein's back. His train sped him toward Berlin, capital of worthless money. A railroad is usually not the best place for serious work; even Max Wertheimer, who dreamed up Gestalt psychology while riding a train, had to disembark to do the actual experiments. However, work's dreamy side was unusually fruitful for Einstein, and a train rocking south from Copenhagen was as good a place as any to think about quanta or unified fields. "The kind of work I do," Einstein once told a friend, "can be done anywhere."

Most colleagues saw their work as an act of discovery, but Einstein identified more with Beethoven than with Columbus. Scientists of previous centuries, he told one audience, "were for the most part convinced that the basic concepts and laws of physics were not in a logical sense free inventions of the human mind, but rather they were derivable by abstraction—that is by a logical process from experiment." Presumably, many people in the audience twisted a little in their seats because they still held to that "old" opinion. Einstein continued, "It was the general theory of relativity that showed in a convincing manner the incorrectness of this view." His most famous theory had been pulled from open sky. The rabbit had not even come from a hat—just out of the clean, blue air.

Speeding around curves, hearing his train's persistent *pahddin-da,* *pahddin-da,* Einstein was poking for still another rabbit. He had liked to base thought experiments about relativity on trains that run at very high speeds along perfectly straight, smooth roadbeds, but the Berlin-bound train's gentle swaying showed that those examples depended on imaginary trains. Relativity had been limited to uniform motion—movement at a constant speed in a single direction—yet real motion is much more higgledy-piggledy. Naturally, Einstein would have liked to include those irregular motions in his theory, but accelerating motion seems to be absolute. Einstein's train had left Copenhagen with a sudden move that sent his chair pressing into his back. The passengers all felt that same pressure and knew the train had begun to move. Nobody on the station platform felt an opposite pressure. There seemed to be nothing relative about the train's sudden accelerating motion. Einstein had been dissatisfied with the limits of his relativity theory, but accelerated motion's absolute nature seemed to disqualify it from a more general law.

At first Einstein had made no serious attempt to extend his theory. Then in 1907 he noticed still another limit to it. Johannes Stark, who in those days was a great Einstein supporter, asked Einstein for a contribution to a journal that he edited. As Einstein began tinkering with possibilities he realized just how much he had united in his relativity theory—mechanics, electromagnetics, optics, the conservation of matter and energy; in fact every law of physics save one seemed to come under relativity's wing. That one exception was the most famous of all physical laws—gravity. Einstein's instinct, of course, was to wonder if he could tug on relativity somewhere to get his law to cover gravity, too. Looking at it quickly—that was another of Einstein's characteristics; he could see at a glance what others realized only after prolonged study—he determined that relativity was hopeless as far as gravity was concerned. Gravity, through a very odd trait, hides crucial information about falling objects.

As is well known, gravity defies common sense by making everything fall at the same speed. People ordinarily expect a 500-pound anvil to fall faster than a 1-ounce jacks ball. Ancient, medieval, and renaissance physics agreed that heavy things fall faster, but Galileo established that everything falls at the same rate. So there is no telling whether a falling object weighs 5 or 5,000 pounds. Relativity's equations demand some way of calculating the weight or energy of a falling object, but the equality of gravity's freefall effect hides that data. *Ergo*, relativity and gravity could not fit together.

Einstein was still working in the Swiss patent office when he wrote the paper for Stark's journal, and he enjoyed ample time to think of other things. Switzerland's taxpayers got their money's worth, however, for he suddenly had what he later described as "the happiest thought of my life." It was not a great step that resolved a long search, but one of those Detective Columbo moments where Einstein understood what he had to prove. It occurred to him that the explanation for gravity was like the solution he had found for electromagnetics. An electromotive force had seemed like something absolute that was created by a particular action, but it turned out to depend on relative systems. Gravity also seemed absolute, but Einstein suddenly saw that it, too, depends on a kind of relative system, an accelerating one.

In an accelerating system, everything accelerates together and so stays together. That idea suddenly transformed freefall's equality into an inevitable and obvious point. In uniform systems, say, on the deck of an aircraft carrier, there is no relation between weight and motion. The airplane sitting on the deck of a carrier moves at the same speed as the pilot standing beside it. That fact is hardly mysterious. In an accelerating system, too, everything moves together. As a train speeds up, seats, luggage, and passengers stay together. The same holds for falling objects in gravity. They "fall" together.

When Einstein recognized this unity between acceleration and gravity, he had no ready examples of people working together in free fall. Today we have all seen film of skydiving teams jumping together from an airplane and falling together. The divers do not need to weigh the same. They can even toss objects to each other, just as people on the deck of a moving ship can play catch. Because they stay together without effort and play catch successfully, the skydivers can think of themselves as being at rest, just as the ballplayers on a cruise ship can think of themselves as at rest. (Granted, the divers will get into trouble if they cling to their idea so resolutely that they fail to open their parachutes.) If one of the skydivers has a radar gun, it will show that the other divers are standing still.

So much of relativity can be observed during a train ride. Clickety-click, on his way between Copenhagen and Berlin, across the German countryside, Einstein could find a wealth of relativistic examples in that simple experience. The pressure surge at startup, the children playing with a ball in one of the compartments, the luggage tag swaying from the suitcase handle, the slender magazine lying still beside a fat passenger, even the light coming in through the window—these ordinary things point toward physics' deepest secrets. Einstein's physics is railroad-age physics.

Outside Einstein's window the train moved through the other Germany, the rural land of towns and villages where Berlin's modern attitudes had no strength. Barely a quarter of Germany's people lived in cities of more than 100,000. The other three-quarters were in smaller settlements, half of them even in villages of 2,000 or fewer. They were the places whose young men had been annihilated during the war, communities whose adults had wept when the emperor fled and whose hope had dissolved into an ocean of worthless money. It was because there were so many of these countrified, traditional Germans, that eventually the Nazi party could be voted into power despite its incorrigible weakness among city voters.

Einstein had no interest in or sympathy for that Germany, and it is difficult to imagine him peering through the window at the farmlands in their fertile glory. He was never a great observer, even though science stories classically open with somebody making a puzzling observation: Arthur Compton notices that X-ray frequencies change as they scatter; Darwin notices that the finches of the Galapagos islands are like South American finches, only different. The scientists then, says the story, struggle for the bathtub moment that will explain their observation. With Einstein, it is not clear that he ever noticed an odd phenomenon in his life. His starting point lay in spotting solutions to contradictions between concepts. Next, he would imagine a possible solution to the discrepancy. Finally, either he saw that the proposed solution would not work or he pressed on to the grand step where he saw how to prove that his new solution was correct.

This peculiarity confounds historians who believe that they must look for some observation that set Einstein rolling. In looking at the original theory of relativity, they struggle to establish that Michelson's and Morley's experiment or Fizeau's experiment were the forces that gave Einstein a push. For relativity's second go-round they like to consider a series of very delicate experiments conducted by a Hungarian baron, Roland Eötvös. The baron had shown that two kinds of masses have identical values. Physicists spoke of *inertial mass,* which is determined by measuring the acceleration of an object when it is hit by a force. A pool hustler takes a billiard ball's inertial mass into account when he "kisses" it firmly enough to make it move, but not so hard that it soars over the table's pocket. A second physical concept is *gravitational mass.* That's the amount you measure when you put an object on a scale. Physicists distinguish between gravitational mass and inertial mass because the masses appear in different equations. Eötvös had shown with careful experiment that despite this theoretical distinction, their values are absolutely equal.

Einstein, it seems, knew nothing of the Eötvös experiments, but the findings fit perfectly with what he called the equivalence principle, the idea that gravity is indistinguishable from acceleration. Einstein's Detective Columbo insight was that the theoretical distinctions between gravitational and inertial masses were heads and tails on the same gold coin. With that idea Einstein abolished the problem of acceleration's absolute nature. It is true that the passengers aboard a train feel a jolt when the train begins to move, while people on the station platform do not, but the relativity lies in the passenger's perspective. There is no way to say absolutely whether the back of the chair is accelerating into the spine of the passenger, or if the passenger is pressing into the chair. If we could say that it was absolutely one or the other, then we could know for an absolute fact that acceleration is pushing the chair forward into the traveler or, again as absolute truth, gravity is pulling the traveler backward into the seat. But there is no way to settle the matter absolutely. Acceleration and gravity are dependent versions of the same thing. When a car suddenly stops, are the passengers pushed forward, or is the car seat pushed back? A seat belt works just as well in either case.

One action that did seem absolutely more powerful than others was the movement of Einstein's imagination once it had grabbed hold of something. While people who are used to thinking of themselves as smart, well above average, and quick-witted are still puzzling to themselves about relative accelerations, Einstein has sped ahead. Relativity in the original theory referred to uniform motion, but now Einstein's attention focused on fields. The equivalence between acceleration and gravity was his concern now. The equivalence of uniform motions appears in special cases where the rate of acceleration equals zero. All of this:

- The equivalence of acceleration and gravity
- The reason everything falls at the same pace
- The unity of inertial and gravitational mass
- The solution to the apparent absoluteness of acceleration
- The shift of theoretical focus from relative motion to relative fields

This whole physics menu was plucked out of thin air while Einstein was daydreaming at his patent-office desk. Mathematicians and novelists routinely turn airy nothing into solid something; physical scientists do it less often.

The experience left Einstein with the optimistic feeling that solutions to mysteries might pop into his head at any movement. Returning from Copenhagen in the summer of 1923, he had been looking 18 years for a clear understanding of radiation, yet, truly, he was not discouraged. The contradiction between wave and particle was like the contradictions he had seen between uniform and accelerating motions, or between mechanical and electromagnetic motions. Light's appearance as sometimes one thing, sometimes another might be just a dependent phenomenon. Understanding of the critical circumstances might be just around the corner.

Ideas behave in a way that is the reverse of how photons behave. You can never tell in which direction a photon will go; with ideas, Einstein had learned he could never predict where one would come from. So he stayed alert and kept all views open. It meant reading all the mail, no matter how nutty, and talking freely to any stranger who was bold enough to approach. The Czech novelist Max Brod remembered with amazement "the ease with which Einstein would, in discussion, experimentally change his point of view, at times tentatively adopting the opposite view and regarding the whole problem from a new and totally changed angle."

He always returned for a fresh look at the quantum. The paper that he sent Stark sketched what an extended theory of relativity would have to prove, but quanta called Einstein back immediately after finishing it and he set relativity aside. Meanwhile his career as an academic physicist finally began to take form. He left the patent office to teach in Bern and later in Zürich. He received his first honorary degree and read his first paper at a science conference. In early spring 1911 he moved with his family to teach at the university in Prague, and during that speedy rise quanta, quanta, quanta held his thoughts.

Part of gravity's lower priority can be explained by its unprovability. Newton's theory was already so accurate that its equation had only one known problem. Mercury's orbit was a little off. All the other planets move as perfectly as God's own clockwork, but Mercury loses time—very little time: 43 seconds of arc per century. Still, the centuries build up. If Ptolemy had used Newton's equation 18 centuries earlier, Mercury would have been 12 minutes of arc out of step by 1907 and the crisis would have been very clear. Most physicists, however, were confident that astronomers would eventually find some simple explanation.

Einstein had also imagined an effect that Newton had wondered about but not grasped. Does gravity bend light? Newton thought it might, but nobody worked out a solution until, in 1801, an astronomer named Soldner gave it a stab. He was ignored, however, because Newton's gravity pulls only things that have mass. Once the wave theory of light took hold, Newton's query and Soldner's speculation seemed impossible. Waves do not have mass. Einstein, however, concluded that light does bend when it enters a gravitational field. As usual with relativity, the prediction is visualizable in terms of the train that Einstein rode between Copenhagen and Berlin. Imagine that while the train sped through the night, a shaft of light passed through the compartment window to flash briefly against the compartment's far wall. The light is no part of the train's relative system, so it will not move with the train. During the time that the light travels from the compartment window to the compartment's wall, the train moves forward a little bit. From Einstein's perspective inside the train, the light beam will not travel in a straight line perfectly parallel to the facing compartment wall. It will bend; the light beam will be a little closer to the back of the train when it hits the wall than when it entered the window. Normally we would say that the light does not really bend. The train has merely moved forward while the light was traveling straight. But that rebuttal is common sense, not relativity theory. From passenger Einstein's relative perspective, he is sitting still. The train has not moved forward, the light's path has just bent.

Well, asks the common-sense thinker, if the train is not moving, why has the light path curved?

Easy, judges Einstein, it bent because of the train's gravitational field.

The train's gravitational field, wonders the common sense thinker. Does the train have a gravitational field?

Sure. Newton showed that anything that has mass has gravity.

So, realized Einstein, light passing through any moving system will appear to bend, and since gravity is equivalent to accelerated motion, light will bend in a gravitational system as well.

This is where many smart people balk and refuse to follow Einstein. Wait, we wonder, can we be so sure that this light bending will work in gravity just because it works for motion? Maybe there are prudent reasons for doubting and hesitating, but Einstein did not doubt, did not hesitate. Like a rock climber hanging over emptiness, he saw the equivalence principle as his only fingerhold, and he was determined to pull himself up as completely and robustly as he could. His theory of gravity would, therefore, assume that light did bend. The trouble was that this bending is almost impossible to detect. We can all

LIGHT BEAM RELATIVE TO ITS SOURCE: Newton's first law of motion holds that an object in motion travels in a straight line unless acted upon by a force. This law can be demonstrated with a flashlight. An observer holding a flashlight sees the light travel in a straight line, even when it enters the compartment of a moving train. Drawing by Pascal Jalabert.

LIGHT BEAM RELATIVE TO A RAILROAD COMPARTMENT: A passenger in a railroad compartment sees light from a flashlight bend slightly as it travels from the window to the wall. From the passenger's perspective, the train is standing still so gravity must have forced the light path out of its straight line. In general relativity, not just time and motion are relative, but also the shape of a line. Drawing by Pascal Jalabert.

agree that in principle light bends as it travels from train window to train wall, but in practice the bend is beyond all discovering. It takes about a hundredth of a nanosecond for the light to travel from window to wall. During that instant a train traveling at 60 miles an hour moves only about three-hundred-thousandths of a millimeter. Testing for such tiny effects looked hopeless.

However, while he was teaching in Prague, Einstein thought of a way to test for these absurdly small distortions. Light passing right by the sun will be bent by the sun's gravity, and by the time the light travels another 93 million miles to reach the earth, even very slight changes will be detectable. Normally, of course, the sun blinds observers to any light passing by, but stars become visible during a solar eclipse when the moon blocks out the sunlight. Einstein even calculated how far starlight should be bent. Assuming that Newton's value for the force of gravity was correct, he found that starlight grazing the sun's border would bend 0.83 seconds of arc. That looked promising, and Einstein took a long break from his quantum efforts.

Typically, Einstein had zigged while everybody else was zagging. He had proselytized for years, trying to enlist the interest of other physicists in the quantum. In autumn 1911, those efforts finally inspired the first Solvay Conference, which moved the quantum onto physics' center stage. Einstein joined Solvay's party and was delighted by the event, but he had switched his thoughts to a revised relativity and he continued to act as a lone mind thinking his way through a maze.

The task, as Einstein saw it, was to develop a theory of gravity that led logically to the equivalence principle, so the theory would have to predict bending light. Newton's theory already worked amazingly well, so Einstein's new theory would have to give Newton's results wherever the old equation worked. The theory would be an extension of Einstein's own theory of relativity, so it would also have to match Einstein's results for cases of uniform motion. In other words, he needed a theory that managed to swallow the two greatest theories in physics; Einstein, a true jumping horse, never shied from an impossible leap.

He did worry, however, that relativity seemed to be coming apart before his eyes. Mathematical theories stretch over the real world rather like fitted sheets over a too-large mattress. When you fit one corner, another pops open. Einstein had begun by looking for a way to cover mechanical motion and electromagnetic motion at the same time. He fitted his theory to cover them, but accelerated motion and gravity were still exposed. Then he saw a way to cover them, but, in response, parts of relativity's corner came open. The trouble lay with his axiom that the speed of light in a vacuum never changes.

Can anything in the real world remain constant? Doesn't much of the pleasure of train travel come from the way passengers can sense all those tiny adjustments and shifts as the train speeds along the rails. It is never as smooth as an escalator's climb. You feel the rhythm of the joints in the rail. The gentle swaying seems to hug you; it provides an almost human reassurance that you are in strong, capable hands. Everything rocks just a bit. Even the luggage tag hanging down from the suitcase handle sways slightly to its own tune as the train carries it along. Isn't nature full of these fluxes? How can anything, even something as grand as light, move along at only one constant speed? Mustn't that be a dream?

Einstein feared it might be. He realized that when light seems to bend it should change speeds while rounding a curve. Horse racing fans can see that the inside track is much shorter than the outside. As thoroughbreds dash along a straight-away the outside runners keep pace with those on the inside, but going into the curve the outside horses fall behind. The curve does not slow down the runners, but the further outside a champion runs, the further it must travel. The inside has the advantage. It works that way for horses and motorcycles, and it should for light, too. Yet experiments have shown that light on the outside of a curve keeps perfect pace with the inside light. To do that, the light on the outside has to be going faster than light on the inside curve. Doesn't that contradict the rule that the speed of light never varies?

By then, Einstein had had plenty of practice at overcoming this kind of challenge; the solution lay in time's relativity. Light rays on the outside curve could keep pace with the inside rays if the inside time went slower than the outside time. To test whether it is the speed of light or the march of time that changes, Einstein proposed to use light's built-in clock, its frequency. Electromagnetic radiation has a specific frequency of some number of cycles per second. That is what the v in the photon's hv represents. If the speed of light changes, it will have no effect on the frequency, but if the time were to change so that the light's time was no longer synchronized with our own, the color of the light would seem to change. Suppose time changes for something with a frequency of 100 cycles per second. If time slows by half, an observer will count only 50 cycles per second. We can then see with our eyes that the light's clock is running slow.

Light frequencies are visible from a range of 750 trillion cycles per second (the violet end of the rainbow) to 430 trillion (the red end). With so many up and down motions per second, a light wave serves as a wonderfully delicate clock. Even slowing time by a trillionth of a second (a picosecond) will produce physical results. If time slows very slightly, light of 500 trillion cycles per second might appear to have only 485 trillion cycles. This is much too subtle a difference for the eye to detect, but perhaps a very precise measuring device could discover the change. If Einstein were right and time does slow down, an outside observer would see light's frequency decrease. Because red has visible light's lowest frequency, this lowering of light's frequency is known as a "red shift." The light does not really become red, but its frequencies move toward the rainbow's red side.

This new notion of time slowing marked a change from Einstein's original idea of relative time. The 1905 theory saw shifting time as a property of the whole relative system. Time flows at the same pace for everything at rest on an aircraft carrier. It might be flowing slightly differently for people on a lighthouse, but those people are in a different relative system. In Einstein's revised theory, time could flow differently for things in the same relative system. They just had to be in different parts of a gravitational field. Time could be ticking at one pace at the top of a building and another pace at the bottom. Time was no longer like a platform that keeps everything together. Instead, picture a river that carries everything in one direction, but where a complex pattern of eddies and winds gradually separate floating objects that started out side by side. A drift in any direction can bring enduring changes in relative position. Time, the universal experience, was becoming a wobbly measuring stick.

The radical heart of Einstein's new thinking about time is caught in the old metaphor, based on Newton's laws, that God was a great watchmaker who set the universe in motion and then let it flow with time's own perfect regularity. Hidden somewhere in that image is the notion of time being as absolute as God Almighty. Now Einstein was saying, no, time is not God, is not absolute. It changes with motion; it changes with gravity.

Einstein traveling from Copenhagen to Berlin found himself like Alice down the rabbit hole. The world that passed his window did not match the one imagined by his fellow passengers. For them the train was moving, the countryside stood still, and there were no two ways about it. Not for Einstein. "Eat me," said a cookie, and Einstein had eaten, discovering as he did so that physics worked just as well if you said the train was at rest and the countryside sped by. Time, too, for passengers outside the rabbit hole, was easy. It flowed at a steady tick-tock, as measured by their pocket watches. "Drink me," said a vial and Einstein drank, concluding thereafter that as he swayed with the railway car he was bobbing and bouncing through a world where time's pace rose and fell like a canoe struck by the wake of a passing tanker. That little swaying luggage tag was bobbing in a time all its own.

Time's wobbly nature did let Einstein tuck one corner of his theory in place, but immediately another corner popped open. His beloved relativity principle began to look untenable. That principle held that the same physical laws are true for every observer. The laws of physics were the same for Einstein on the train and for people in the countryside farms. When Einstein boarded a railway ferry to cross from Denmark's islands to the European mainland, the physical laws on the ferry matched the laws on the train and in the farmhouses. By this principle, Einstein meant that if you performed experiments in those separate locations you would discover the same laws. If you were Einstein, you could discover relativity anywhere. Thus, an Einstein aboard a railway ferry could perform an experiment, do a calculation, and say, "You know, that phenomenon looks different from the train." Thanks to the math, the experimenter could tell everybody aboard the ferry exactly how things looked from the train. But with wobbly time these calculations were not working. You could not be sure what the other people saw.

Even patient Einstein appears to have grown frustrated with the way one corner or another of this theory kept coming undone. We can guess at this upset partly because it would rattle anybody, although Einstein's patience and optimism made him unlike just anybody. A more telling sign of his deepening frustration was the old flag that seems to have been impressed on his psyche, he began a passionate love affair. During this period, he visited Berlin on an unsuccessful job hunt. Wherever Einstein went, a woman or two was likely to catch his eye. Usually these incidents implied no more emotion than a sailor's tryst in the latest port of call, but physics had grown unsatisfactory and, in Berlin, Einstein began an affair with his recently divorced cousin. Stymied by gravity, he began writing indiscreetly to Elsa.

April 30, 1912:

My dear Elsa, No sooner had I left you than the thought started to weigh on me that it would be impossible to write to you since you are so closely watched. How happy I was then today when I saw from your letter that you found a way that will allow us to stay in touch with each other. How dear of you not to be too proud to communicate with me in such a way. I can't even begin to tell you how fond I have become of you during these few days. . . . In your amiable way you are making fun of me, but this does not make me like you even one iota less. I am in seventh heaven when I think of our trip to Wannsee.

In Prague, a colleague and musical friend named George Pick told Einstein to expand his mathematical horizons; he needed a new language more than he needed a new experiment or even a new idea.

Einstein, however, was not that kind of poet. Shakespeare was liable to reshape English every time he dipped his quill. Proust, to express his psychology of time, developed a new style of French. Newton, when he found the existing mathematics was insufficient for analyzing motion, invented differential calculus. Einstein called that deed, "perhaps the greatest intellectual stride that it has ever been granted to any man to make." Einstein's working language was also mathematics, but in this area he was no Newton. Although Einstein was surely the most original physicist ever, his mathematical talents lay simply in applying known methods to his physics.

May 7, 1912:

I suffer very much because I'm not allowed to love truly, to love a woman who I can only look at. I suffer even more than you, because you suffer only for what you do not have.

Einstein had outrun Euclid's geometry. During his trip to Japan, Einstein told an audience, "Describing the physical laws without reference to geometry is similar to describing our thought without words. We need words in order to express ourselves." But sometimes we need new words. There were other geometries besides Euclid's, although most physicists would have said that switching to some other geometry was absurd. Euclid's was the true geometry; anything else was a mathematician's game. But Einstein had been much impressed by Poincaré's argument that Euclid's geometry is no more the true geometry than meters and kilograms comprise the "true" system of measurements. Geometries are tools for describing reality.

Johannes Kepler once, as he struggled with the problem of defining Mars's orbit, wrote an equation that expressed changes in the planet's position. Today any mathematician or astronomer would look at that equation and know immediately that it described an ellipse, but in Kepler's day geometry had not yet been combined with algebra; he had no way of knowing what his equation meant. Einstein in Prague was stifled that same way. His physics demanded more geometry than he had.

In midsummer 1912, he left Prague to teach again in Zürich. As soon as he was back in Switzerland he looked up his old classmate and friend, Marcel Grossman, who had become a mathematics professor. Einstein poured out his troubles just as, eight years earlier, he had revealed all his frustrations to Besso. "Grossman, you must help me or I will go crazy," Einstein begged.

Grossman did come to the rescue, acquainting his friend with a geometry developed 60 years earlier at Göttingen by Bernhard Riemann. Perhaps the system's most notable feature is that it has no parallel lines. Space in Riemann's system is often said to be "curved," but that is a Euclidean prejudice, like saying Arabic books run "backward." Riemann's geometry has straight lines, and Einstein's physics still accepts Newton's axiom that an object in motion will continue to move in a straight line. It is just that straight lines in the Riemann system do not look straight from Euclid's perspective. In Riemann's system, "bending" can turn out to be a Euclidean illusion. Light passing through an accelerating or gravitational system might appear to bend in Euclid's space, but not in Riemann's. Einstein did not yet understand how that could be physically real, but he knew that if he could prove it, he would have solved his case.

March 23, 1913:

What wouldn't I give to be able to spend a few days with you, but without . . . my cross! Will you be away all of (August until the beginning of October)? If not, then I would like to come for a short visit. At that time my colleagues will not be there, so we would be undisturbed.

To understand how a straight line can seem crooked, use a ruler and a colored felt pen to draw a line on a flat sheet of transparent plastic wrap. Next, crumple the wrap into a ball. That's the whole of the experiment. Using a ruler to draw a line assures that the resulting mark meets the classic definition of a straight line; it is the shortest distance between two points. Crumpling the wrap does not stop the line from being the shortest route between any two points it crosses; however, crumpling the wrap did change the line's appearance. Viewed from outside, the line looks crooked, yet a creature living on the wrap's surface would find that the line still marked the shortest path to all its points. So is the line really crooked or straight? There is no absolute answer.

Einstein's abandonment of Euclid changed the problem facing physics. Newton looked at the sky with Greek eyes and said the planets do not move in a straight line. There must be some force pulling or pushing the planet out of straight motion. With a little detective work, Newton concluded the force was a pull. This pull was mysterious, acting instantly at a distance, but the facts of elliptical orbits were plain. So Newton's gravitational equation tells how to calculate gravity's pull.

Einstein argued that the bending of lines was relative. From the perspective of anything traveling along the line, it goes straight. A light beam entering a moving train compartment goes straight, from the perspective of the light beam, but from the passenger's view it bends. This relativistic conception of paths demanded a new way of describing how the wrap crumpled, so to speak—that is, what the space looks like to a particular observer. The equation that Einstein wanted would have to describe the apparent curving and crumpling of space.

Surely there were tourists, perhaps many of them, aboard the train from Copenhagen. How exciting for them to discover a bonus, a world-famous passenger having coffee and a smoke in the dining car. With the mark turned worthless, vacationers could visit Germany and use their real money to buy anything they wished. The Germans saw them and resented the invasion. They had not been warned about how humiliating it is to lose a war.

Tullio Levi-Civita and Gregorio Ricci, two Italian mathematicians, had developed a calculus based on Riemann's geometry. Grossman began to study and teach Einstein the system. Einstein, of course, was an able pupil and moved swiftly to geometry's front-line, where he began chopping himself a path. In 1913, Einstein and Grossman published a paper together in which they offered their new equation, but Einstein was unsatisfied. The proposed new law appeared to ignore the requirement that a single set of circumstances must always lead to the same result. Without that regularity, causality disappears from science.

Newton's gravitational equation was the perfect example of mathematical causality. Insert the same numbers and you always get the same result. But Einstein's equation seemed to allow for an infinite variety of outcomes, all based on the same gravitational circumstances. That variability did not sound right, so Einstein kept working.

February 1914:

Dear Else, Don't be angry with me for my being such a poor letter writer. This does not mean that I love you any less. I cannot find the time to write because I am occupied with truly great things. Day and night I rack my brain in an effort to penetrate more deeply into the things that I gradually discovered in the past two years and that represent an unprecedented advance in the fundamental problems of physics.

He moved from Zürich to Berlin but continued rethinking gravity. The World War began; he kept working. The work would have been impossibly lonely if Einstein had not been so enchanted by his own questions. A pacifist in a wartime capital, he was hated by many. Many others resented the way he silently laughed at their anguish over the world's distaste for German grandeur. A theorist trying to replace science's most successful theory, he found little sympathy for rethinking Newton. Yet, as 1915 began he wrote a friend, "I firmly believe that the road taken is in principle the correct one and that [people] later will wonder about the great resistance the idea of general relativity is now encountering." Einstein's colleagues repeatedly saw his efforts as mere windmill jousting. Then, when he bagged another giant, they could only gape in amazement.

Some of Einstein's optimistic tone might have been bluff, yet there is something fantastic about even bluff optimism in January 1915. That winter was one of the most dismaying in world history. In the preceding four months, the western front had formed. A trench line had been dug from the English Channel to the Swiss border and each side was besieging the other. The war was going to persist indefinitely. In Munich, Thomas Mann had been working on *The Magic Mountain* since July, 1913, but by early 1915 he found that the war was too distracting and he set his novel aside for the duration. But Einstein was able to set mass slaughter aside, just as he put love aside, to concentrate on his physics.

———◆—◆———

In midsummer 1915, Einstein underwent a crisis of faith. He realized that his previous papers on gravity formed "a chain of false steps" and that he was clinging to untenable ideas. At first this recognition was as frightening as finding oneself lost in a wood, but he screwed up his courage and found the strength to take yet another grand step. He had been insisting that relative systems are as real as the deck of a ship, or the platform of a lighthouse, or some such concrete place. Being real, relative systems are limited in number and meaning. Truly, he realized, they are merely arbitrary abstractions, something theory imposes on the cosmos in order to provide a reference point. You can make your reference point the bridge of a ship or a spot 10 feet above the ship. You can align your reference coordinates so that they run through the center of the earth, or tilt them to run some other way. Whatever you choose, it is just a reference point and not part of nature itself.

Einstein's abandonment of real, relative systems might have been his most difficult feat of imaginative resourcefulness. Relative systems had been crucial to his original theory; now he threw their reality aside. It was the kind of back flip that Lorentz had missed when he held to the reality of absolute time and thus lost the meaning of his own equations. And, as happened 10 years earlier when Einstein took his great step toward relativity, he followed the rejection of real reference points with a month of frenzied work, tracing the mathematics and logic of his new understanding. In the two years since he had written the paper with Grossman, Einstein had tried many new equations and calculations. Now he tossed all that aside and went back to the mathematics of that 1913 paper. "In two weeks," Einstein recalled, "the correct equations appeared in front of me!"

When the outpouring was over, Einstein wrote to Sommerfeld, "During the last month I experienced one of the most exciting and exacting times of my life." Exciting it was, as clarity and proof spilled over his notebooks. The problem of an infinite number of solutions to the same set of circumstances simply evaporated. There were an infinite number of ways to define the reference points, but each definition led to a single solution. A predictable outcome—causality—had returned to his physics.

Factual support for the theory appeared as well. The old equations that had proven so useful in the original theory of relativity could be derived from the new equations. They served in a special case—instances of uniform motion through Euclidean space—and Einstein began to call his first theory "special relativity." Then he was able to show that in all the cases where Newton's gravitational equation worked properly, his own results were so similar as to be identical.

Next he turned his attention to the case where Newton's equation did not work so perfectly, Mercury. Solving his equation for the orbit of Mercury was demanding work, and when the solution matched reality Einstein felt palpitations of the heart.

"For a few days," he told Ehrenfest, "I was beside myself with joyous excitement."

"Something actually snapped," he told another friend. Ideas pulled from his head had matched the data gathered by sweating astronomers winking through telescopes.

If at that moment reporters from the world press had appeared at his doorway and Einstein had become at once the world's most famous scientist, he might have understood why he was so applauded. But he received no such cheers. Every Saturday for four weeks in November 1915, he rose from his seat in the Prussian Academy to report his progress. On November 4, he described his break with absolute coordinates. November 11, more mathematics. November 18, still more mathematics, demonstrating now that the equations settled the problem of Mercury's orbit and that he calculated that the bending of light passing by the sun would be twice the size of the prediction he had based on Newton's gravitational value. November 25, he presented his final equation and reported that "any physical theory that obeys special relativity can be incorporated into the general theory of relativity." His work on relativity was done.

A few days later he told Sommerfeld that the final equation he presented was almost identical to one that he and Grossman had considered two years earlier but rejected. It had contradicted the conservation of energy. Einstein had changed the equation just a little bit, such a little bit that it hardly seemed to matter except that it now conserved all the energy. And it was just that little bit that led to the correct calculation of Mercury's orbit and to the revised figure for the bending of starlight passing the sun.

Sommerfeld responded that he was astonished to hear Einstein claim to have solved gravity. "You will be convinced once you have studied it," Einstein replied, "Therefore I am not going to defend it with a single word."

But unadulterated happiness never lasted long in Einstein's heart. He soon noticed that a corner of his mattress that had seemed tucked tightly had again become visible. Physics suddenly had two fields—one gravitational, one electrical—sitting right on top of one another with no connection between them. It was as frustrating for Einstein as mind-body dualism is for more ordinary thinkers. There must be a way, he was sure, for one law to account for both of those fields. Almost eight full years later, as Einstein's train returned him to Berlin, he was still looking for that way.

18

Intellectual Drunkenness

S tepping down from the train into Berlin's Lehrte Station, Einstein was back in catastrophe's homeland. Sweden had been confident, Denmark assured, but agonized Germany was walled off from the world of postwar normalcy and wanted someone to blame. Its money had become worse than worthless. Any counterfeiter who transformed a blank sheet of paper into a million German marks would have been reducing the paper's value. Once Einstein would have needed a separate valise just to carry the money to pay the porters who carried his bags, but Germany had responded to the hyperinflation by printing bills of enormous denominations, so the famous scenes of Germans carting wheelbarrow loads of money just to do their grocery shopping were no longer necessary. Indeed, the train station still looked clean and prosperous. That summer Hemingway reported from Germany that tourists, even those who traveled extensively, saw no suffering. "There are no beggars. No horrible examples on view. No visible famine sufferers nor hungry children that besiege the railway stations." So Einstein was able to make his way home, without having to step over pitiable neighbors.

"The tourist leaves Germany wondering what all this starving business is about," Hemingway added, "The country looks prosperous. On the contrary in Naples he has seen crowds of ragged beggars, sore-eyed children, a hungry-looking horde. Tourists see the professional beggars, but they do not see the amateur starvers. For every ten professional beggars in Italy there are a hundred amateur starvers in Germany. An amateur starver does not starve in public. On the contrary no one knows the amateur is starving until they find him. They usually find him in bed."

But the wall between Germany's ruin and the rest of the world's indifference was not quite an iron curtain. The money rot had begun to spread. In the capitals of the victorious powers, economics ministers were alarmed. The value of the French franc was sinking and the British pound had started to wobble.—*Good!* was the normal, suffering Berliner's reaction. *Let them find out what we have been suffering.*— *Good!* thought the more conspiratorially minded. *Now they will taste what they have done to us.*—*Good!* cursed the bitterest cynics. *Now that it is threatening them, they will put an end to it.*

Who were *them* and *they*? The Jews took some of the blame, but the French probably took more. The war had been over for almost five years, but neither the French nor the Belgians seemed to have lost any of their fury or instinct for revenge. They had sent troops deep into Germany to seize and hold the Rhine Valley and were agitating to create a separate Rhine republic, splintering Germany into smaller units of despair. "Germany will wait in vain for us to vacillate even a moment," France's prime minister declared during the dedication of a war memorial, "France will walk this road to its end." Unable to fight back, Germans quarreled among themselves about how to explain their continuing agony. God's punishment. An allied conspiracy. Jewish thievery. Socialist bungling. Capitalist greed.

Most explanations for the catastrophe required some kind of leap. On the one side towered the unprecedented, unimaginable destruction of money and everything money supports. On the other sat the fully precedented, frequently imagined things that one fears. The money catastrophe served as a kind of hieroglyph in which readers saw the name of their own personal dread, and everybody had one. Einstein tended to believe more in humanity's general folly than in conspiracies and he blamed the chaos on Germany's soul rather than on foreigners or internal plotting. "Germany had the misfortune of becoming poisoned," he wrote in a note to himself, "first because of plenty, and then because of want."

Einstein was one of the very few in Germany who had crossed the line, gone into France, examined the remains of the western front, and returned, deeply shaken, to Berlin. That visit, a year before his journey to Scandinavia, had shown him just how high the wall between Germany and victorious Europe still stood. Einstein had been cheered in America and Britain, but his reception in France was mixed. France's most distinguished physicists welcomed him, but he decided against speaking at the French Academy because some of its members threatened a walkout if a German tried to speak before them. The Society of French Physicists also snubbed him, but Madame Curie showed her support by attending his lecture at the College de France and the prime minister, Paul Painlevé, who had also been a professor of mathematics, ushered at the door.

A French physicist, Henri Bouasse, an expert on acoustics and hydrodynamics, announced that "the French spirit, with its desire for Latin lucidity, will never understand the theory of relativity." Remarkably, Bouasse did not think he was slandering France, "It is a product of the Teutonic tendency toward mystical speculation."

But the other side was just as visible. On the morning he left France, Einstein got into a car with three distinguished Frenchmen—Maurice Solovine, Paul Langevin, and Charles Nordmann. Solovine was a friend from the patent-office days in Switzerland, a mathematician and Einstein's French translator. Langevin was a leading physicist and expert on magnetism. Einstein later said that Langevin would have surely discovered special relativity for himself, "had that not been done elsewhere." Nordmann's title was astronomer of the Paris Observatory. He had written a wonderful little book about relativity. All three citizens had accomplished what Bouasse proclaimed impossible for Frenchmen. They understood relativity.

The countryside they passed through between Paris and the German border had been murdered. It was not just that the ground had been torn open by artillery shells, or that the houses were blasted apart, or that chunks of human bone seeded the fields. Even the trees had been gassed to death. Their roots still propped them up, but the branches had been barren for years. In Reims, the heart of Champagne country, they stopped for lunch in a restaurant where two high ranking French officers were dining. This area had been the front itself, with French forces entrenched in and around Reims, while Germans held the high ground. The region had endured a steady pummeling over the four years of war, with some of their heaviest dueling going on while Einstein was in his creative frenzy, solving general relativity. The Reims cathedral, one of the symbols of traditional France, where many kings had been crowned, was now a ruin. The city's homes were as devastated as the dwelling places of the ancient Trojans.

In the restaurant, the officers recognized Einstein at once and sent glances his way. None in Einstein's party could be sure what would happen. The world's most famous German was seated among supposedly enemy veterans and surrounded by devastated lands and towns. But they ate without disruption. Einstein's trio of escorts were puzzled to learn their guest was such a teetotaler that even in Reims he refused to down a glass of Champagne. "I do not need wine," he explained, "because my brain is acquainted with intellectual drunkenness." They stood to leave the officers and their woman companion rose, too, and then bowed toward the celebrated physicist-pacifist. So even among veterans there could be respect for Einstein as a man and as a champion of reason.

But the wall between the two societies was still very strong. After leaving the restaurant, Einstein's party continued eastward, crossing the ground that had been the no man's land between trench lines. Even with the machine guns and artillery removed, those no-man's barriers lived in many hearts. Trade between the once-enemy neighbors had still barely resumed. Before the war, France had bought about 8 percent of Germany's exports. In 1923, with Germany's export trade slashed by 40 percent, France was buying barely 1 percent of that output. Trafficking in ideas and courtesies was weaker still.

The wall also blocked Einstein's professional reach. He had ready news of the ideas coming from Göttingen and Copenhagen, where his photons were viewed without sympathy and where research focused on other matters. Remarkably, it was in Paris that interest in the meaning of hv continued. The seeds had been sown at the great Solvay

Conference of 1911. One of the conference secretaries was Maurice de Broglie, an experimental physicist based in Paris. He returned from that first Solvay as an enthusiast for Einstein's light quanta, although, as an experimentalist, he had no clear way of theorizing about them. The theorist in Paris who was most energetic about trying to understand wave-particle duality was Marcel Brillouin, father of Leon, the physicist who worked to maintain the French-German wall by demanding to exclude Einstein, "whatever his genius," from international meetings.

Paris's work had no influence on Göttingen's side of the wall and could not have seemed more remote if it had been carried out in Nairobi or Bengal. Yet it was in Paris that in September, 1923—while members of the Göttingen-Copenhagen alliance stiffened themselves for their final dual over light particles—that physics at last began to grapple with the mystery of wave-particle duality.

Maurice de Broglie's younger brother, Louis, had been greatly impressed by Maurice's enthusiasm after Solvay 1911. Louis had planned to study humanist subjects in preparation for a career in the French civil service, but his brother's work and the writings of Poincaré persuaded him to take up physics. He quickly showed a keen interest in and appreciation for the toolkit of ideas, equations, and techniques that Einstein had given physicists.

Louis was working on his doctoral paper in September when he used Einstein's toolkit to gain a Detective Columbo hunch about what must be proved if the quantum mystery was ever to be explained. Instinctively, de Broglie concentrated on the experimental evidence that most clearly illustrated the wave-particle paradox. De Broglie hoped for a "synthetic theory of radiation," that is, a theory that combined waves and particles into one thing. This hunt for the nature of light kept getting more complicated. Ancients had thought light might be something in us, a ray sent out from the eye, but a medieval Arab scholar known in the west as Alhazen studied pinholes of light and their ability to project images against a dark wall. He concluded that light did not come from the eye but consisted of particles moving in straight lines. Newton, too, studied single beams of light and came to a similar conclusion. Then Young did his two-slit experiments showing the interference patterns typical of waves. Maxwell's optics ended all dispute. Anything Alhazen and Newton studied could be explained by waves.

But now we have a complicating explanation in which light again acts like a particle to produce the photoelectric effect and Compton effect. Set up a two-slit experiment as described by Young to show the wave interference, but have the light land on a metal wall. Now the wave interference will be visible on the metal, but at the same time the light will knock out electrons in the manner of particles and produce a photoelectric effect. Here you have light moving like a wave yet interacting like a particle. It is the photon's primal mystery.

De Broglie attacked this puzzle by imagining a photon as a kind of clock. Alarm clocks are solid things, but they have some sort of internal ticking that makes them different from solids like soccer balls and bricks. The photon is bricklike enough to carry momentum and produce the Compton effect, but something about it is always ticking. Just what might be acting so regularly, de Broglie could not guess.

Reaching again into Einstein's toolkit, de Broglie pulled out Einstein's two most famous equations: $E = mc^2$ and $E = h\nu$. To the untrained eye, the E's in both of these equations look identical, but the physicist knows better. The first E refers to energy available in a particle of matter (mc^2). An electron, for example, can somehow be transformed into pure energy, consuming the electron in the process. We can call that the material E. The second E defines radiant energy, the energy carried by a massless quantum of light $(h\nu)$. That word "massless" is important here. Mathematically it means that mass (m) equals zero. So if you use mc^2 to calculate how much material energy you can get from a photon you find that $m = 0$; so $E = 0c^2$, or just plain old nothing.

De Broglie, of course, knew of these differences, but he combined the two equations anyway, saying $mc^2 = h\nu$, material energy equals radiant energy. This was as bold a union as Einstein's assertion that inertial mass and gravitational mass were physically equivalent. Einstein's bravado had ultimately yielded new insights into both gravity and acceleration. If de Broglie's match-up proved correct, physicists could anticipate insights into particles with masses, like electrons, as well as into those without masses, like photons. If material energy (in electrons) was equivalent to radiant energy, then perhaps electrons, like photons, were waves as well as particles. If so, then it should be possible to shoot a beam of electrons through two slits and see an interference pattern. De Broglie's theory, then, was subject to experimental testing.

De Broglie reached yet again into Einstein's toolkit. The most versatile of Einstein's yardsticks was, of course, relativity and its ability to compute the way things look to different observers. Using his combined Einstein equation, de Broglie derived two equations for a particle's internal clockwork. One figured the clockwork relative to the particle's own reference frame, while the other defined the clockwork from an outside perspective. De Broglie found what Einstein had already shown: thanks to the peculiarities of time's relativity, the outside observer sees time slow down. Einstein had described the process in his theory that gravity produces a "red shift," but the red shift is a feature of radiation. By combining Einstein's two great equations, de Broglie was implying that the same rules apply to atomic particles. It is not just photons that are going to show a red shift; electrons should show it, too.

At this point in the story the difference between electron and photon has started to grow a little hazy. Electrons have mass, photons do not. That is something, but by emphasizing the differences, we miss what de Broglie was beginning to suspect—the great similarities between atomic particles and photons. De Broglie used the word "mobile" for both particle-like waves and wave-like particles. The ambiguous term let him speak of electrons and photons together without worrying about whether they were waves or particles.

De Broglie focused his effort on the difference in a mobile's frequency as seen by the mobile itself and the frequency as seen by an outside observer. The mobile's measure of frequency differs from the outsider's measure, but remember, we are talking about the same mobile. The frequencies are relative, just as speed is relative or pathways are relative. Looking for something that could link the two relative frequencies, de Broglie imagined "a fictitious wave." Anybody who knew that Einstein had been thinking about "ghost waves" might rise to attention here. De Broglie suggested that this fictitious wave moving with the mobile is in harmony—"in phase," to use the technical jargon—with both the particle's view and the outside observer's view. He concluded that "this harmony of phase will always persist," and then he generalized in a manner worthy of Einstein that "any moving body may be accompanied by a wave" and it is impossible to distinguish between "motion of body and propagation of wave."

In de Broglie's mobile, wave-motion and body-motion are joined together like lovers on an opera house stage. They remain individual yet are united through the harmony of their music. The difference between de Broglie's wave and Einstein's hard-sought ghost wave is that de Broglie's is like a duet, while Einstein had been looking for a separate item, something like a bullwhip, that could carry a wave but did not have two-ness built into its nature.

Duets, of course, can join more than two singers in harmony. The instruments in the orchestra pit play their parts as well, and all these sounds join together into something that, for all its richness, can be reduced to a single groove on a record. The tenor's voice, the soprano's sweetness, the violin's contribution along with those of the cello and oboe are distinct to the ear and yet they all combine to describe one solitary line on an old recording disc. De Broglie imagined that the harmonies of many photons could also be described by a single wave, which he called a phase wave.

Einstein had missed this musicality of the quanta, and as he so much loved music, it is natural to wonder why he missed it. Einstein's *skree-skree*, however, was not harmonious, and he had no sympathy for teams that acted in unison. His strength lay in the statistical effects of individuals acting individually—of, say, water molecules crashing independently against isolated specks of dust. Even as a child, unified action held no appeal for him. Watching a military parade in Munich as a boy, he was unmoved by the music, the color, and the thousand men tramping as one. Horrified, he told his parents, "When I grow up I don't want to be one of those poor people."

De Broglie, however, hoped even the photon's primal mystery might become comprehensible if he thought of a light beam as a phase wave of photons moving in step. Suppose someone behind a curtain throws an alarm clock. We cannot see the clock move behind the curtain; however, if our ears are keen enough, we can hear it move. Each tick will sound from a different place. If we take the ticking sound as the whole story, we will become confused. The clock will seem to jump between different points without going through the space between, but if we consider this ticking as a clue to the clock's general activity we can begin to make sense of it: Something moving gives us a regular report—but not a continuous one—on its where-abouts.

De Broglie thought light could be considered similarly. It is moving continuously through space but can interact with matter only when the elements of its phase wave are in harmony. Those harmonious moments mark when the light's presence can be discovered, just as ticking sounds periodically reveal a clock behind a curtain.

If de Broglie's phase wave was to survive, it had to explain not only the mystery of wave-particle duality, but Alhazen's pinhole image and Young's interference. There are two standard techniques for analyzing light motion. One is geometric, drawing lines to show how light travels or is bent by a lens or reflected by a mirror. Alhazen had used that method to argue that light is a particle, so de Broglie's problem was to explain how it could be created by a mobile. De Broglie's hunch was that the light moved in a continuous wave while the light's "ticking" occurs at the same point on each wave. Suppose an alarm clock rides a small roller-coaster that carries the clock up and down along a continuous track. It takes one second to cross one full roller-coaster hill and the clock only ticks once a second, so the clock will always be at the same point on each hill when it ticks and a person tracking the motion by the sound will suppose that the clock is moving along a straight line.

There is an old military joke about a sergeant drilling recruits. He has them parading back and forth, but then he makes a mistake. He is too slow in giving an order to turn and the squad marches straight into a wall. Not flummoxed for even a second, the sergeant barks, "If you were real soldiers, you would have all hit that wall together." The same principle applies to photons hitting the wall of a pinhole camera. Only those that are in step when they hit the wall will "tick" and reveal their presence, so only evidence of straight line motion will be observable.

Young's two-slit experiment reversed de Broglie's problem. Alhazen's geometric optics were no help here. Wave dynamics were needed to account for this interference. Could de Broglie's mobile also explain it? This time it was not quite so easy. He had to add a few extra hypotheses to his account (something Einstein in particular was always loath to do), but the essence was basic. If the slits are sufficiently tiny, passage through them disrupts the light, bending it so that it scatters a bit, just as soldiers marching through an obstacle will be disrupted. Trained soldiers will reform their group and resume marching, but of course if some soldiers cross one obstacle and some cross a different obstacle, the separated groups will no longer be in step with one another. A physicist would say they have gotten out of phase. And, of course, photons are not trained soldiers who can alertly keep together. Howsoever the slit distorted their paths, that is how the photons travel, and when they hit the wall, some will be "ticking" and reveal their presence while others will be out of phase and interact with nothing. The interference pattern on the screen does not reveal where light is absent and where it strikes; it only shows where "ticking" light strikes. And that pattern matches perfectly with the dynamic analysis of how waves will interfere with one another. De Broglie was naturally happy to have found in the interaction of light and matter a "fundamental bond that unites the two great principles of geometrical optics and dynamics."

De Broglie had created an agenda for quantum physicists. Instead of trying to explain away the troubling paradox of wave-particle duality, physics should weave it into the heart of any explanation of how matter and energy interact. The wave theory was good, but it placed too much emphasis on continuity and missed out on the tick-tock way that mobiles interact with the larger world. Particle theory was good too, but it was overly mechanical and assumed that every particle interacts with everything it touches. If a particle hits a wall, it makes a bang. So if the wall is not banging, there cannot be any particles hitting it. This dogma missed the way mobiles can strike a wall between ticks and not interact at all. The new physics would have to be a science of discontinuous interactions between mobiles that move continuously, or at least it would if de Broglie's hunch proved right.

Like any poet or sculptor, de Broglie thought his creation too beautiful to be false, but others might want harder evidence. In particular they might want proof that electrons are really mobiles and can create interference patterns when fired through narrow enough slits. De Broglie was cautious enough to close the final paper of his miracle September with a voice of controlled excitement, "Many of these ideas may be criticized and perhaps reformed, but it seems no little doubt should remain of the existence of light quanta. Moreover if our opinions are received, as they are grounded in the relativity of time, all of the enormous experimental evidence of the 'quantum' will turn in favor of Einstein's conceptions."

A closing like that would normally have come to Einstein's attention immediately, but the wall between Paris and Berlin had put the two cities out of phase and they no longer interacted coherently. In Paris, de Broglie's paper was seen as remarkable, but probably crazy. In Berlin, people were concentrating on matters like the price of milk, which one fine morning stood at 15 million marks and the next day had jumped to 30 million.

19

The Observant Executrix

In October, shortly after Louis de Broglie's wondrous September, newspapers in Berlin began leaking Einstein's plans to visit Russia. It was news that was bound to irritate the anti-communist middle class, and many of the intellectual elite, too. The universities were still drifting rightward. At the University of Berlin, the new rector had erected a war memorial to honor the students killed in combat. The Latin inscription read, *Invictis victi victuri*: to the undefeated and to the defeated who will emerge victorious. It was hardly the time for Einstein to boost his local popularity by going east.

Bolshevik Russia, in the autumn of 1923, was perhaps the only place in Europe where the mass of people suffered more than in Germany, although even Russia's financial situation was more stable than Germany's. Lenin had replaced confiscation with a "new economic policy" that returned Soviet living standards to Russia's prewar levels; however, the country's only strong institutions—the Communist Party and the Red Army—depended exclusively on force and fear. All previously known bases of political legitimacy—religion, family, democracy, and even money—had been replaced by fur-hatted commissars who studied the peasants for any sign of a stray opinion.

If the fur hats had any claim to authority, it came from Lenin's prestige, but the sands of that boast were about to run out. During the previous 12 months, Lenin had been felled by two strokes. Recently, on the 10th of October, he had astonished his bootlickers by climbing out of bed and insisting that he be taken to the Kremlin. The return proved his force of will, but his physical strength was only a memory; the next day he was back in bed, where he stayed. The strokes had deprived him even of the ability to shout "Revolution." "Rev . . . rev . . . rev . . . vo . . . vo . . . vo . . . lu," was the best he could do. Adding to the desperation of the moment was the absence of any clear successor. In the background, of course, Stalin was already moving his dependents into place, but that was behind the curtain. The Soviet Marxists boasted that their scientific history was as inevitable and preordained as a solar eclipse, but they had not yet found—and never would find—a way to manage their succession in a manner that suggested an inevitable historical process.

Why Einstein would pick that crisis hour to travel eastward to the Bolshevik state was impossible to say. He had been invited before but always refused. He tilted at different windmills from the ones Marxists favored. Immediately after the revolution, some enthusiastic socialists had gone to Moscow for a look, but Einstein was never that red and besides, those early days were far past. A tour in late 1923 by a celebrity of Einstein's stature was a real coup for communism, or it would have been if the tour were real. The news of Einstein's plans and visit, however, was a fabrication, an early illustration of German nationalism's Big Lie tactic.

Several rightist Berlin papers carried reports of Einstein's scheduled visit and, as happens with successful propaganda, the news spread to other outlets. Eventually, one bold liar even gave an account of Einstein's arrival in Saint Petersburg and of his plans to stay there for three days of looking and admiring. Berliners were used to news of their hero-scientist traveling, but the papers that started these new reports of Russian travel were mostly nationalist, anti-Jew publications that normally shunned accounts of Einstein's honors. They ran these travel tales to discredit a man they loathed. Meanwhile, the real Albert Einstein spent a quiet October in Berlin, doing nothing to deny the stories, not even after his own occasional outlet to the German public, the *Berliner Tageblatt*, reprinted the rumors. He had responded very energetically and diligently to the first set of anti-relativity, anti-Semitic lies that followed the industrialization of his fame, but after four years of hearing his name beaten on the drums he had learned the impossibility of keeping a celebrity's record straight. Fame had turned out to be as much slide show as spotlight, leaving people free to project whatever image they chose onto his form. Even before his celebrity, Einstein had a tendency to stand apart from the world and watch its follies with amusement; by 1923 he almost never spoke in his own defense.

Yet even autonomous Einstein could not live forever detached from all society. When October became November he traveled to Leiden to visit his friend Paul Ehrenfest. It was not a perfect escape. Whenever Einstein went outside, strangers approached to say Hello, to stand beside him while someone snapped a photograph, or even to interview him for some newspaper or magazine. Despite this hubbub, the change was good for the Berliner. Ehrenfest lived his homelife at the antipodes of Einstein's style. It was filled with the kind of love that sets its own expectations. His wife, Tatiana, was a trained mathematician who worked alongside her husband as collaborator-in-chief. One of his sons, Vassik, had Down syndrome, or, as it was known in those days, mongolism. Vassik was being raised at home and brought constant worry to his parents, who wondered how he could survive without them. Einstein could never have stood to have a wife who was an equal, or children who rested so firmly at the center of need and attention, but he took pleasure in the Ehrenfest household. Host and guest sat across from one another, conversing energetically and enthusiastically about Compton's discovery and the photon's sudden status as fact. Of de Broglie's ideas, naturally, they said nothing, for they remained in perfect ignorance of them.

Leiden had none of the Bohemian bustle that characterized Berlin during even its most wretched days. It was a quiet university town surrounded by canals, flatlands, and windmills. Einstein's pleasure there foreshadowed the easy way, 10 years later, he would take to Princeton, New Jersey, another university town with quiet neighbors. Leiden was 1,000 years old, and mostly those years had gone unnoticed beyond the town's borders. There had been a sudden spasm of involvement in history in the quarter-century before and after 1600, when rebels opened the dykes and flooded Leiden's countryside to drive out Spanish troops. Leiden then became a center, perhaps *the* center, of free thought with a free press. The Elsevir family settled there and published Galileo when his books were banned in Italy. English Puritans also published books there, before packing their presses and becoming the American "Pilgrims" of Plymouth, Massachusetts. Rembrandt was born in Leiden in 1606, enrolled in Leiden University at age 14, and began his life as a painter there, but in 1631 he moved on to Amsterdam, putting an end to Leiden's half-century as a shooting star.

Three hundred years later, the Leiden that Einstein visited was still a center of free thought. The Ehrenfests had landed there because they were uncloseted atheists who could not find a job in most of Europe's universities. It was hard enough being Jewish, but religious scoffers besides! Yet Leiden, and really all of the Netherlands, was proud of the Ehrenfest presence. Besides, most of physics was being done in backwaters that season. Yes, de Broglie was hard at work in cosmopolitan Paris turning his bloom of ideas into a doctoral thesis, but the rest of physics puttered along in less boisterous locales—St. Louis, Copenhagen, Göttingen.

Compton's work had restored photons to the top of the agenda, but nobody knew how to seize the hour. Bohr still sought an argument that could explain away the Compton effect. Einstein was eager to move on, but a paper he produced that December showed he still did not yet know how to advance. Max Born summarized the general state of theorists when he wrote Einstein, "As always, I am thinking hopelessly about the quantum theory." Yet Einstein remained as optimistic as ever, optimistic about both his own abilities and the likelihood of solution. "Optimistic" is indeed too tame a word for his confident faith that ultimately natural events have discoverable, natural explanations. Maybe he would find it, maybe somebody else would, but Einstein never questioned that nature had its reasons and that some fox would catch the reason's scent.

Ehrenfest did not expect that he would ever be the fox. He even worried about his ability to follow the progress of his colleagues, though he shared Einstein's assumption that quantum effects had their reasons. "He's a skeptical fellow," Einstein had said admiringly of

Ehrenfest, meaning that his companion liked to see plenty of evidence before embracing an explanation. He did not mean that Ehrenfest was moved by the deeper philosophical skepticism of a philosopher like David Hume, who doubted that we could derive truth from sensory experience. In 1923, there were not many westerners who believed in neither God nor scientific law.

The mystery about those two friends, sitting comfortably at the table in a Dutch kitchen, slicing a piece of cheese and chatting about Arthur Compton was the way they seemed so ordinary and yet something made them quite extraordinary. Artists, philosophers—crafters in every form of imagination—have always sat in cafes, inns, houses, and kitchens, to exchange words. A scene like Einstein conversing with Ehrenfest has been a staple of intellectual history since the time of the Babylonians. The scientific turn of mind that hopes to learn about nature has been part of every civilization, just as every civilization produces its poets and musicians. And yet something new must have been added, because in recent centuries, those following, say, 1600, this kind of conversational give-and-take had been rewarded to an extent previously unimaginable. That was the mystery of Einstein and Ehrenfest together. What was the extra ingredient in their discourse that led science to such success?

That same day, frustrated colonials in distant places like Dakar, Delhi, Peking, and Cairo spoke heatedly in kitchens and cafes and demanded to know how millions of intelligent, capable people stayed under the dominion of European power. The folly of the World War had shown that the Europeans were not gods, yet the victorious nations persisted in their blasphemous arrogance. It was difficult to rank traits and determine which ones gave these occupiers their power. Was it their Christianity, their derby hats, their science? Whatever it was, angry, ambitious youths wanted to find it, to make their way, and wipe that smirk off those arrogant white faces. It only rarely occurred to anyone in these disrupted societies that it was exactly their arrogant confidence that lay behind the newcomers' power.

Many of the conquered peoples had scientific traditions that were older than the Europeans'. France and England had just carved up the Arab world, yet it was the Arabs who had once rekindled learning in

Europe. India's millions were under the thumb of a few thousand British soldiers and civil servants, but India's mathematical and engineering feats predated the Romans. The Chinese squirmed under a series of unequal treaties and lamented in vain that their traditions of scholarship were thousands of years older than their barbaric masters'. An eavesdropper in Leiden, hovering quietly over Einstein and Ehrenfest, might have eventually sniffed out the unspoken doctrine that distinguished them from the sages of Alexandria and Uxmal. Like the classical Greeks, they were masters of mathematical technique. Like the Chinese, they could cite close observations of fact. And like the Arabs they believed that something lay behind the facts. But while the Arabs believed the something behind nature was God's inscrutable will, Einstein and Ehrenfest believed that nature was lawful and the law was ultimately knowable. The difference between the Einstein and Ehrenfest table-talk and that of the other scientific traditions was that those other traditions had been content to formulate laws of experience. Einstein's tradition looked for something deeper, a law stamped on the universe that forces it to act as it does.

Where such a belief came from, how it was foreshadowed in the poetry of Dante and Chaucer—all that is another story, but by 1600 the writings of Francis Bacon and Galileo Galilei showed that the change in the minds of a few had taken hold. Looking back on that age, Einstein once mused about Galileo's contemporary, Kepler: "How great must his faith in the existence of natural law have been to give him the strength to devote decades of hard patient work ... entirely on his own, supported by no one and understood by few." Originally this faith had been accompanied by traditional Christian beliefs, but over time faith in nature's order had proven subversive of orthodox creeds; science had come to seem anti-religious. Boston's Cardinal O'Connell would tell an audience that Einstein's theories "cloaked the ghastly apparition of atheism ... [and] befogged speculation, producing universal doubt about God and his Creation." But modern science had emerged from Christianity, just as surely as Christianity had sprung from Judaism.

Ancient science had a different paternity. The first Greek philosopher was Thales of Miletus; he might just as well be remembered as the first Greek scientist, for his ambition appears to have been discovering how nature really works. Even in those times, that sort of ambition always seemed to bring a little authority with it. The Nile priests knew what day of the year it was, and thus when it was time to plant in preparation for the river's flood. So others depended on the priests. Dependence brought power. It did not, however, subvert religion.

Thales's knowledge, too, brought power. Herodotus told the story of a battle in 585 B.C. between Greeks and two Asian tribes. Both sides had been stalemated for years; they often fought but never managed a decision. Once they even engaged in a night battle, very unusual in those days, but still the war persisted without promise of conclusion. Eventually, however, the Greeks entered a battle armed with a secret prophecy. Thales foretold that an eclipse of the sun would occur during the forthcoming engagement. The battle began. The sky went dark. The Greeks fought on. The Asian warriors were too startled by the sun's disappearance to maintain their positions. At last the Greeks could press to a decision and settle the war in their favor.

It was a promising beginning for a people who would eventually conquer the eastern Mediterranean, but for all their learning the ancients had only what Einstein called "laws of experience." Even Ptolemy's astronomy, which was based on Euclid and remained consistently logical, did not dig down deeper for natural law; that is, it did not search for an order behind the surface behavior. The Alexandrian Greeks were a people who through genius and toil assembled bits and pieces of a jigsaw puzzle, but who never figured out that the snippets were parts of a whole. Without that extra idea, they had no reason to believe they could complete the picture.

Einstein's conversations with Ehrenfest in the Leiden kitchen, however, were part of a common effort to explain quantum effects in terms of the rest of nature. Neither man would have considered for a moment that quanta did not obey some natural law, yet before modern science, opinion had been divided only between pious believers who saw God's will determining outcomes according to some inscrutable plan and the cynical realists who agreed with La Rochefoucauld's maxim that "Chance and caprice rule the world." Neither side had any reason to expect that a general rule of natural behavior might exist. Thus, when they saw some oddity in nature, they looked for fudges rather than for new generalizations. The planets usually travel around the earth from east to west, but every so often one of them moves west to east. Moderns, even those of us with no scientific training, immediately wonder why the change. What is going on in nature to make that reversal? To us it seems an inevitable question, but in other societies it was not so routine. The Roman and medieval astronomers looked for a mathematical workaround rather than a natural principle. By training and culture they sought an additional complexity in the mathematics of planetary motion, not a new simplification. Their breakthroughs tended to be mathematical rather than meaningful, as when the Alexandrian Eratosthenes suddenly conceived of a way to calculate the earth's circumference. His insight was brilliant, marvelous and achieved through pure geometry. It added a fact to the world's body of hard-won knowledge, but because it explained absolutely nothing about nature it was the kind of achievement that never interested Einstein. Before Copernicus, most scientific milestones consisted of that kind of nonexplanatory achievement.

The modern belief in the possibility of coherent, natural explanation had allowed science to transform the world and gave its disciples undreamt-of new powers. They held the authority of Thales and the Nile priests cubed, then squared again. The idea that nature was, as Galileo put it, "the observant executrix of God's orders," released a butterfly from a caterpillar's cocoon. The old science had crawled; now it flew and gave believers a new project—understanding the world as God had made it. This idea that people could discover nature's laws was the most profoundly revolutionary doctrine since the prophecies of Mohammed, and it changed the earth's social face just as quickly. Lore and craft that had been stable for centuries, or even millennia, were suddenly transformed. Alchemy became chemistry. Astronomy divorced astrology and began learning new things. Geology blossomed from the rules of thumb that guided mining and quarrying. Biology and meteorology condensed from the void.

Einstein was the latest in the string of prophets who saw nature in the modern, Galilean way, and who had used his faith in nature's lawfulness to expound a series of unexpected, profound revelations about how the universe moved. Other believers were not as talented, but they still had an expectation of natural orderliness that gave their talk and their meetings a special quality, even when chatter or assembly came seemingly to naught. In the spring of 1924, for instance, the triennial Solvay Conference on physics took place without dropping any fruit. That was the meeting whose isolationism had prompted Einstein to tell Lorentz he was not coming and did not want to be invited to any future such conferences. Solvay 1924 showed plainly that physics had become stuck.

But even when stuck, modern science shows its strength. Classical astronomy was stuck for 1400 years between Ptolemy and Copernicus, but nobody at Solvay 1924 thought such prolonged idleness was possible again. When today's scientists are lost, they cast about for ideas. They go over what they know. They try little changes; they test radical hunches. That's what Einstein did when he was alone. It's what Einstein and Ehrenfest did when they were together. And it is what went on at Solvay in 1924. Confident of nature's lawful, discoverable behavior, they were like the Greeks armed with Thales's prediction. They pressed on after others stopped struggling. The chatter that spring in Brussels seemed idle and limited, but much more damaging was its exclusion of nationally incorrect savants. Two-thirds of the Solvay 1924 attendees came from victorious countries, one-third from neutral lands, and zero-thirds from the defeated Central Powers. Nobody attended from institutions in Germany, Austria, Hungary, or Prague. Planck, Sommerfeld, Max Born, Wien, Heisenberg, and Pauli were barred from the door. Bohr could have attended but did not bother. So, when in the course of social events, Paul Langevin told of his student de Broglie's ideas, there was nobody ready to seize the story. The hot quantum topic that season was Bohr's BKS theory. And yet every reader knows that with modern science's relentless push toward understanding, somebody like Einstein was bound eventually to take note of de Broglie's ignored idea.

20

It Might Look Crazy

Not quite two months after 1924's Solvay Conference, Einstein received a letter from a young, unknown Hindu named Satyendra Bose, a physics teacher at the University of Dacca. Writing in English, Bose included a manuscript. Anybody near the public eye occasionally receives unsolicited letters that introduce a manuscript. Einstein, in a perpetual spotlight, got them all the time. Many people feel overwhelmed by far less of this sort of material than was piled up on Einstein's desk, but he read through his mountain of stuff and did not turn away just because the contents were unorthodox.

Bose's letter bore several marks of the crank and had a secret trait besides. Bose did not mention that his manuscript had already been rejected by one respected journal, *Philosophical Magazine*. The rejection is not surprising, because, although the article presented a statistical theory, it violated several rules of statistical thinking. Bose—and this is also typical of crackpot authors—did not realize how far he had strayed from proper reasoning and later conceded, "I had no idea that what I had done was really novel. . . . I was not a statistician to the extent of really knowing that I was doing something which was really different." That part about "I was not a statistician" is another trait of the crank. Crackpots typically are not well acquainted with the field they have encroached and do not have the training to appreciate their own radicality.

Just as cranky as the manuscript's secret history and unknowing heresy was the letter-writer's outrageous sense of entitlement. "I do not know sufficient German to translate the paper. If you think the paper worth publication I shall be grateful if you arrange for its publication in *Zeitschrift für Physik*." Grateful indeed. Bose added from his outpost on the floodplains beneath the Himalayan wall, "Though a complete stranger to you, I do not feel any hesitation in making such a request. Because we are all your pupils though profiting only by your teaching through your writings." And in that remark Bose expressed what leads all these strangers to contact and make demands on the more-famous. Modern communications has deluded the anonymous millions into feeling a relationship between themselves and a celebrity. Einstein recognized the delusion, yet remained *mensch* enough to take these letters seriously and to study them.

Bose's letter did include a mathematical reference that might have encouraged Einstein's reading. It used one of those mathematical formulae that make the untrained want to turn their eyes away, as though exposed to a painful light. "I have ventured to deduce the coefficient $8\pi\nu^2/c^3$," Bose wrote, tipping the fact that he had learned some real physics somewhere. So Einstein read the manuscript and then began translating it into German. Neither incompetent nor crazy, Bose's paper was the first revolutionary contribution to quantum theory in a dozen years, and if Bose had not realized how revolutionary he was, that was also part of the quantum tradition. Planck had had no inkling of the kind of trouble he was setting off when he first proposed the quantum.

Bose's paper—like Louis de Broglie's still unknown work—began by taking Einstein's photon seriously. He proposed that radiation be thought of as a kind of gas. Possibly the *Philosophical Magazine's* reader stopped right there. Gases, after all, have mass and are made of molecules scattered loosely across space. Photons have neither mass nor molecules. Einstein kept going.

"Your ideas are interesting," Langevin had told Louis de Broglie when he had proposed a similar gas-radiation analogy, "but your gas has nothing to do with true light." De Broglie rarely could get a hearing for his ideas. Einstein was more respectful with Bose, agreeing that the challenge lay in getting beyond the simple gas-radiation analogy to something physically meaningful.

Technically, Bose's novelty lay in the way he derived Planck's equation. Planck's work imagined waves emerging from oscillators (like bedsprings); Bose's reasoning assumed radiation particles in the manner of Einstein's photons, but these particles had several odd qualities. First, Bose's photons are as temporary as the waves that washed over the foundering Titanic. Classical particles cannot be destroyed, although waves can be absorbed and disappear. Thus, with the "non-conservation of photons," wave-particle duality begins to take on mathematical form. Bose's particles were also indistinguishable from one another. This detail, too, gives mathematical expression to the mystery of wave-particle duality. Classical particles can always be distinguished from one another, at least by differences in position. Waves, however, are spread out over time and space and, as everybody's home stereo system shows, two or more waves can be in the same place at the same time. This mathematical union of wave and particle properties was real progress beyond the bald, dumbfounding statement that experiments show light to be both particle and wave. Einstein recognized that achievement, even if Bose himself was not quite sure of it.

Einstein's English was not perfect, but he went to work that summer translating. When he told his secretary *cum* mistress that he was abandoning her to pursue happiness in the stars, the Bose statistics were the immediate stars of his attention. While Bohr, Heisenberg, and the rest of the Copenhagen-Göttingen axis were still looking for a way to wriggle out from under the Compton effect, Einstein was a mile further along the road, examining the photon's mathematical nature.

Einstein had spent years looking for the step that would make sense of photons, and then it turned out that this puzzle was so subtle that it required Bose and Einstein to dance together in a two-step trot. Bose's step dealt with a quantum's radiation side while Einstein advanced its material side. Bose provided a natural explanation for Planck's 25-year-old quantum equation. In Bose's picture, light particles come pouring out of their source like clowns out of a circus car. It seems impossible for so many to fit into anything so tiny. Photon indistinguishability allows them to press together in a way unseen in the material world, where two objects cannot be crammed into the same place at the same time. Freed from one of nature's normal limits, a radiation "gas" can flow in densities forbidden to atomic and molecular gases.

Einstein could never look at such an idea without tinkering. He saw the formal similarity between Bose's radiation gas and a real gas, and—while the Bose paper was still waiting for publication—Einstein went before his favorite intellectual club, the Prussian Academy, to once again read aloud a startling paper that shifted the ground of physics. He told the members about Bose's work and what it meant about ideal gases. In this first paper he introduced a modification into Bose that recognized the most notable difference between photons and true particles: the way photons come and go while particles last forever. The atoms in the wire of a photographer's flash, for example, were cooked billions of years earlier in some lost star, but the photons that the wire emits are born and absorbed in less than a second. Einstein adjusted Bose's equation to reflect that difference between particle and photon, transforming Bose statistics into Bose-Einstein statistics.

He tinkered some more and, thanks to our knowledge of de Broglie, we can see where he was headed. For decades, Einstein and the occasional disciple had been looking for particle properties in light waves. The photoelectric law, Compton effect, and Bose's statistics had settled that hunt. The new idea, in both Paris and Berlin, was to look for wave properties in particles. Einstein wanted to get beyond a formal, mathematical analogy between gas and radiation, to a complete, physically coherent linkage. This was, of course, the same Einstein who had believed that atoms had to be real, and quanta had to be real, and the equivalency principle had to be real; so naturally Bose's gas had to be real as well.

Many years later, Einstein dismissed the Bose-Einstein work as "by the way," and it was a distraction from his interest in a unified field theory, but only somebody of Einstein's standing could say such a thing. It was like Shakespeare downplaying *Othello* as a distraction from his real work on *King Lear*. You can see his point, but goodness gracious! and it was hardly fair to Bose, whose career had just reached its apex.

Bose's step, his explanation of nature in terms of nature, was so much at the heart of scientific achievement that it is easy to overlook how rare it is. Many scientists pass their whole careers without taking such a stride. Bose never again managed such a step, but these rare insights are what makes science so persuasive, and so unlike other rational fields—philosophy, criticism, theology, law, politics, and so on—in which explanations provoke division. Bose had kept his eye on the target, explaining photon action in terms of photon properties without ever jumping to some other subject. Many people, of course, like those explanatory jumps and cannot conceive of an explanation that omits them, considering as no explanation at all anything that leaves out God's will, or subconscious motivation, or class interests, or what-you-wish. That might have been science's most remarkable trait—its attempt to explain nature without changing the subject.

By refusing to jump to some other terminology, science never moves outside nature. Many people, probably even most people, cannot believe that nature is the whole story. "The whole modern conception of the world is founded on the illusion that the so-called laws of nature are the explanations of natural phenomena," the philosopher Ludwig Wittgenstein complained in his notebook while he was posted to the Austrian front lines, "Thus people today stop at the laws of nature, treating them as something inviolable, just as God and Fate were treated in past ages." Einstein, however, looked only to nature, and when he was, so to speak, folding nature into itself—showing, say, that radiation and matter really were bound by the same laws—he was as happy as his heart allowed.

For some reason, or perhaps for no reason at all, it could have been entirely by chance, that season saw a stirring and climaxing of many great efforts to find a way to truth. Besides Einstein's statistical portrait of photons, the bankers had restored the money to order; the lunatic right had been quieted; Hitler was in jail. Thomas Mann had a frenzied month (September, 1924) in which he finally finished his *Magic Moun-*

tain. The publisher then had an equally frenzied time of it and at the end of November sent copies of the thousand-page novel into bookstores. Dada metamorphosed as well when André Breton wrote *The Surrealist Manifesto,* which described an expression "free from any control by the reason and of any aesthetic or moral preoccupation."

While he was working on the physical meaning of Bose's gas analogy, Einstein received a letter from Paul Langevin in Paris telling him about de Broglie's work. Langevin was undecided about accepting de Broglie's thesis and, because so much of it was founded on Einstein's ideas, Langevin wondered what Einstein thought about the manuscript. Einstein read de Broglie's thesis and discovered that, in some ways, his latest work had been foreshadowed. At once, Einstein became an advocate for de Broglie. He urged Langevin to accept the thesis, and he began recommending the document to others.

"Read it," Einstein urged Max Born, "even though it might look crazy, it is absolutely solid." Einstein's behavior in this matter was so absolutely right and decent that we can forget how tempting it is for established scientists, when they find themselves headed off at the pass, to use their own prestige to claim to have been there first. Johannes Stark was always making that sort of claim. Even a giant like Newton was perpetually embroiled in battles over priority to an idea. Einstein seems to have been beyond that style of temptation, and in his second paper on Bose-Einstein statistics he added a footnote urging the world of physics to pay attention to de Broglie.

That second paper was read before the Prussian Academy in early 1925 and startled its audience with predictions of undiscovered behavior by gases. When matter approaches a temperature of absolute zero, said Einstein, its atoms will become as indistinguishable as photons and a new kind of quantum gas will replace the classically observed one. As usual with Einstein's theoretical spectaculars, he was far ahead of the experimentalists. The first laboratory confirmation would not appear until 1928, and it took almost 70 more years to demonstrate that molecules really can condense into a gas with indistinguishable parts. Einstein's paper also showed that his statistical approach to matter and de Broglie's wave approach corresponded exactly, so the two theories snapped into place like the two sides of an arch.

Shortly afterward, Einstein read the Academy still a third paper about quantum statistics. This time he settled the question of whether the third law of thermodynamics applies to gases. The third law states that as temperatures approach absolute zero, the loss of usable energy (the entropy) also approaches zero. According to classical statistics, gases contradict this law, but Einstein showed that the law holds under the new quantum statistics.

Einstein's papers on quantum statistics appeared at the same time that German physics was awash with rumors of laboratory confirmation of the Compton effect and disproof of its rival BKS theory. It was a time of great professional triumph for Einstein. His quantum theories now had a clear factual basis in photoelectricity and X-ray scattering, and a coherent explanatory basis in quantum statistics. Among physicists this achievement was quickly recognized and admired, but it contributed nothing to the Einstein legend, which by then was fixed in hardened cement and now focused much more on the person than on his work.

That spring Einstein toured South America, visiting Buenos Aires, Rio de Janeiro, and Montevideo. He was hailed, of course, and praised for relativity and his antiwar politics. Einstein always thought it odd that ordinary people cared anything about relativity, so he was not puzzled that his quantum work went ignored. The enthusiasm over his politics was equally bizarre, because Einstein well knew that the masses of people are not antiwar. A reminder of that reticence over peace came in Buenos Aires when a local reporter proved to be an old acquaintance from Berlin named Otto Buek. Einstein and Buek had formed half of a group of four Berliners who dared to sign a "Manifesto to Europeans" that protested the outbreak of the Great War. As Buek recalled, they "had overestimated the courage and integrity of German professors." Yet the Germans of 1925 boasted of their Einstein and in Buenos Aires the German ambassador spoke enthusiastically about him. Einstein laughed to himself that although Germans commonly saw him as "a foul smelling flower" they stuck him in their "buttonholes." Isn't life preposterous? Einstein always thought so.

21

Taking Nothing Solemnly

In early 1924, before he sent his article to Einstein, Bose had applied to Dacca university for a grant to spend two years studying in Europe. The school officials took their time considering such an expensive idea. Apart from the cost of two years living at European prices, Bose would need travel money and a supplementary grant to ensure a living for his wife and family. Any officialdom worthy of its stamp pads can delay indefinitely a costly proposal like that, but the cavalry suddenly arrived in the unexpected form of a postcard written in Einstein's own hand. "The smartest man in the world" had written from Berlin to an obscure physics teacher (not even the head of his department) in Bengal to say that he (Einstein) considered Bose's work to be an important contribution to quantum physics, and that he would see that the article was published. Good-bye, bureaucratic delay. Dacca University promptly voted a generous stipend. The German consulate, after one look at Einstein's postcard, gave Bose an immediate visa and waived the standard fee.

Bose then traveled by railway across India's breadth and set sail from Bombay for the wonders of Europe. Arriving in Paris several months after his article had appeared, he discovered that his name was already known to France's leading physicists. Maurice de Broglie put him to work in his lab, and Bose stayed in Paris for almost a year. That was the Paris of expatriate fame. William Faulkner was there that year, living the bohemian life with no money. Aleksandr Kerensky, the revolutionary democrat whom Lenin had chased into exile, had landed in Paris, too. It was the city where Stravinsky composed, Picasso painted, and Gertrude Stein collected.

In physics it was the home of Madame Curie and Paul Langevin. Bose met both of them and he corresponded with Einstein, letting him know that he was in Europe and hoped to meet him in Berlin. Einstein encouraged Bose to come and talk. Yet Bose hesitated to move on. Accounts of Bose's travels tend to express puzzlement over his long stay in Paris before pushing on to Einstein. Perhaps those head-scratchers should read Hemingway's *A Moveable Feast* to get a sense of the pleasures and excitement that held Bose in France. He finally arrived in Berlin a year later, in October 1925, only to find that Einstein was away on his annual visit to Leiden, so Bose had several weeks on his own to look about and explore the German capital.

The city he found was much different from the tormented *stadt* of two years earlier. The hyperinflation had become a bad memory, and even the war promised finally to recede into history. An armistice had ended the fighting in 1918 but had resolved nothing else, not even the military blockade. The Treaty of Versailles ended the blockade but kept the hatred at full boil. Even Germany's borders had remained unsettled. Now, in the fall of 1925, another agreement, called the Treaty of Locarno, was negotiated, this time settling Germany's western borders with France and Belgium. Einstein hoped this new "spirit of Locarno" meant that Germany would at last be welcomed into the League of Nations.

Clearly too, in this recovered world, Berlin would become almost as grand as Paris. Bose found in Berlin a cosmopolis that had already become a greater center of modern art than it had ever been before. The grand talents that were once scattered over many German-speaking cities and regions had begun moving into Berlin, the way geniuses of the Renaissance had swarmed on Rome. Berthold Brecht had been Munich's leading playwright, but he was now in Berlin. Heinrich Mann (brother of Thomas Mann), long considered Germany's finest prose writer, had also left Munich for the new center. Arnold

Schönberg moved there from Vienna, bringing the most radical new music with him. Alban Bern was there too, rehearsing his opera *Wozzeck* for its world premiere in December. Young artists whose careers would span the remaining century were in town. Yehudi Menuhin played the violin while, around the block, Vladimir Horowitz performed on piano.

And Berliners appreciated what they had. "I loved the rapid, quick-witted reply of the Berlin woman above everything, the keen, clear reaction of the Berlin audience in the theater, in the cabaret, on the street and in the café, that taking-nothing-solemnly yet taking-seriously of things," recalled a film critic of that era. Bose absorbed the sights and sounds until Einstein's return from the Netherlands. When the two men at last came face to face, Einstein expected to meet a theoretical physicist and greeted him as such.

"How did you discover your method of deriving Planck's formula," Einstein wondered.

"Well," Bose replied, "I recognized the logical contradictions in both Planck's and your own derivations of the formula and I applied the statistics in my own way."

Einstein moved in to challenge Bose. Did he think that these statistics meant something new about the interaction of quanta? And could he work out the details of those interactions?

With that question, Einstein had brought Bose to the frontier of quantum physics. The great need was for a coherent, intelligible theory that would describe how quantum mobiles, to use de Broglie's term, acted on one another. In particular, of course, Einstein would have welcomed any insights useful in predicting the direction taken by a newborn photon.

Bose, however, had no notions on extending his statistics toward a quantum mechanics. He was, in fact, not really a practicing scientist of the sort Einstein usually dealt with. By vocation and imagination, Bose was a great teacher. He did not strive to understand nature so much as he struggled for ways to help his students understand the achievements of others. He had devised his new way of deriving Planck's formula because both Planck's and Einstein's derivations seemed to him to mingle classic and quantum ideas as injudiciously as a bar-

tender mixing whiskeys. As a teacher, Bose could not ask his students
to pretend to understand something that was fundamentally illogical,
so he had looked for a better explanation that would satisfy his most
alert students. That effort had given Bose's letter to Einstein its odd
tone of crackpot supplicant. Bose had written to Einstein as one
teacher to another ("we are all your pupils") while Einstein had responded as one physicist to another. In Berlin, however, it became
apparent that the two men could not really help one another directly.
Although he taught many people, Einstein lacked a teacher's ambition
to convert people to a way of thinking, while Bose, despite his enduring contribution to theoretical physics, was not driven by a scientific
imagination. Einstein did give Bose letters of introduction and commendation that allowed him to enter all science doors in Berlin and
Göttingen, but the two men did not become collaborators.

In a world where teaching is devalued, even viewed with contempt, this turn of events is seen to reflect ill on Bose. Even Bose
himself spoke rather slightingly of his career, saying, "On my return to
India I wrote some papers. I did something on statistics and then again
on relativity theory, a sort of mixture, a medley. They were not so
important. I was not really *in* science any more. I was like a comet, a
comet which came once and never returned again." Yet in India he
continued teaching, inspiring generations of students through the legend of his meeting with Einstein, and in the way he clarified physics
by demanding that the students see an idea's rough spots and find their
way through it. "Never accept an idea as long as you yourself are not
satisfied with its consistency and the logical structure on which the
concepts are based. Study the masters. These are the people who have
made significant contributions to the subject. Lesser authorities clev-
erly bypass the difficult points."

So the unexpected voice from India was not going to rescue Einstein
from the puzzle patch that physics had become. In December, following his encounter with Bose, Einstein returned to the Netherlands to

join in celebrating the 50th anniversary of Lorentz's doctorate. Europe's leading theorists came to honor one of the founders of their field. Ehrenfest served as host. Bohr had come down from Copenhagen. Madame Curie and Arthur Eddington were there too, along with dignitaries from outside the scientific world. Queen Wilhelmina attended the ceremony and presented Lorentz with a medal.

The biggest surprise in physics during Lorentz's 50 years as a doctor of philosophy had surely been the ruin of perhaps the oldest idea in scientific materialism—the belief that matter was ultimately composed of indivisible little particles that moved exactly the way larger particles moved. The giants celebrating Lorentz's career, and the guest of honor himself, had put an end to that notion. Lorentz had predicted and coined the name "electron" for a particle smaller than an atom that somehow was part of an atom. The initial assumption, of course, had been that electrons themselves were tiny indivisible particles that moved according to the laws of classical mechanics.

Bohr, smoking his pipe and enjoying the Lorentz celebration, had shown that electron motions did not match perfectly with classical ones. And Einstein, with de Broglie and Bose, had recently brought the story even further, showing that electrons are not pure particles but some sort of wave-particle duality. When Max Born (absent from the festival because he was taking advantage of American lecture fees) organized a meeting to discuss de Broglie's paper, one of his students proposed an experiment to test the idea that electrons showed wave properties. "Not necessary," said Max Born's colleague, James Franck. He reported that an American researcher at AT&T had already performed experiments that gave de Broglie's predicted results.

The world was not as the materialists had expected, but how could it be? Materialists had not anticipated the existence of a second fundamental reality: energy. If you reduced all matter to indivisible particles, you still would not have accounted for the energy running all those steam and internal-combustion engines that powered the industrial world. There had to be more to the story than the classical atoms and their endless, random motions.

Einstein had shown in 1905 that the story was going to become irretrievably strange. In the spring of that year he had proved that atoms were real, and in the fall he concluded that energy and matter are different forms of the same thing. Matter can be metamorphosed into energy. The old image of atoms as solid specks of grit was not going to work. Inevitably you were going to find that when you got down to fundamentals—down to apparently indivisible specks like photons and electrons—you were going to meet ambiguities. Specks of energy had particle-like natures, while specks of matter had wave-like properties.

Lorentz, cheered and congratulated, had kept abreast of those changes and proved his continuing nimble-mindedness by remaining knowledgeable, articulate and inventive about the many new ideas that had appeared since he had become *Dr.* Lorentz. The relativity of time, the quantum of action, the electron jump—he had absorbed them all, while staying faithful to the basic scientific ambition of understanding natural phenomena in nature's own terms. For the young and middle-aged physicists at the celebration, this demand for perpetual agility was the promise and challenge of a scientific career. If they could remain as fleet-footed as Lorentz, they could keep moving toward the pot at the rainbow's end where a full understanding of nature lay.

An example of the demands and rewards of science was visible at the celebration itself. Ehrenfest showed off two of his bright students, who had just discovered a third property of electrons. Besides mass and charge, electrons have a "spin." They rotate around themselves the way the earth does. It was typical of Ehrenfest's two sides—great teacher, neurotic self-doubter—that he was producing precocious students even while he worried that the new physics was eluding him.

Bohr was dubious of this latest idea. How does an electron's magnetic field survive the picture of a spinning electron?

Elementary my dear Bohr, Ehrenfest told him. This was the basic stuff of relativity. A rotating electron is, from the electron's point of view, standing still while its electric field appears to rotate about it. And when you have a rotating electric field you have a rotating magnetic one as well.

Ah, it clicked for Bohr, and he saw that physics had taken another stride.

That roomful of savants—talking gossip, talking science, disputing and finding common footing—had characterized science throughout Lorentz's career. It had the look of an enterprise that would go on that way forever. In fact, however, this proved the last gathering of the old scientific order. Lorentz's celebration was like the funeral of England's Edward VII on the eve of the Great War. Kings and potentates had turned out to show the strength of the international system, but really they were bidding it farewell.

22

How Much More Gratifying

Einstein's second paper about Bose-Einstein statistics, the one in which he cited Louis de Broglie's work, sent ripples through the world of avant-garde physics. In Göttingen, it led to the realization that de Broglie's prediction of electron waves had already been observed; yet that recognition led to no further study. The physicists of Göttingen did not seize the moment to look more deeply. In Munich, Arnold Sommerfeld raised the news of de Broglie's thesis with his graduate students. They looked briefly but did not take the ideas seriously. It is often said that Einstein was viewed among physicists as a god and that even the greatest of them would grow humble and silent when he entered their presence. On some level, that is true, but it cannot be the full story because the history of Einstein is stuffed with incidents where Einstein's hints are ignored. By itself, his authority was rarely enough to make a physicist think harder.

Only in Zürich did his colleagues say that they should try harder to understand the baffling Frenchman whom Einstein had seen fit to endorse. Two physicists in Zürich—Peter Debye and Erwin Schrödinger—together held parts of the job that had been offered years earlier to Einstein. The Yes-No answer that Einstein had tendered the Swiss led to his rejection of a coprofessorship at the University of Zürich and the Polytechnic. In his place, Debye had come to the university while Schrödinger came to teach at the Polytechnic. Schrödinger was particularly eager to study de Broglie's ideas, in part because he was already experienced in waves and statistics, and partly because, for Schrödinger, Einstein really was a god. Relativity, especially general relativity, was, according to a future colleague, "his chief love." And like Einstein, Schrödinger thought that the existence of two different physical fields, gravitational and electromagnetic, occupying the same space at the same time was too ugly to endure, and he, too, longed for a unified field theory. So when his hero wrote that de Broglie's ideas "involve more than merely an analogy," Schrödinger decided to study this new theory, according to which "a moving corpuscle is nothing but the foam on a wave of radiation in the basic substratum of the universe." Debye and Schrödinger agreed that Schrödinger would investigate the "undulatory" idea more fully and report his findings to a colloquium.

The problem with de Broglie's waves was similar to the popular joke about the economist who recognizes a fact as being true in practice, but who doubts that it can be true in theory. De Broglie's waves had been observed in the lab, but no theory described them. If there were waves in practice, there should be a wave equation in theory. Schrödinger would have to find the equation that defined nature's wavy groove.

Any such equation that Schrödinger proposed would have to support the well-established laws of motion for large particles like planets orbiting the sun, or balls on a soccer field, or even pollen dust floating on water. In this regard, Schrödinger was like Einstein approaching general relativity. Einstein's equation had needed to embrace gravity, relativity, and the new business about acceleration and light. Schrödinger expected that his equation would have to capture Newton's laws of motion, plus special relativity, plus the established laws of quanta and electron function, plus the new ideas about particles surfing the waves of reality.

Whew! One begins to remember just what it was about Einstein's ideas that made his colleagues grow silent. They were not too crazy, just too hard. Professional scientists cannot risk years of work by running down a likely dead end. Einstein spoke contemptuously of the kind of physicist "who looks for the thinnest spot in a board and then drills as many holes as possible through it," but of course that approach is eminently practical if you are concerned about your career.

Perhaps there was something in Switzerland's mountain air, but Schrödinger, like Einstein, seems to have been a bold jumping horse. He gave himself to optimism and went hunting for a new wave equation. He was already well acquainted with the way sound waves function, and in particular he understood how wave groups or packets act together. That experience gave a path for approaching de Broglie's mobiles. De Broglie's paper had suggested a reason for the most arbitrary feature of an electron orbit—its use of fixed (or "quantized") orbits.

Bohr had visualized electrons orbiting an atom's nucleus, but if electrons really moved through the electric field of a nucleus, like planets orbiting the sun, the electrons would radiate energy and fall into the nucleus like a comet tumbling into the sun. Plainly, electrons do not fall that way. Why not? We know that before moving things can come to a halt they must lose their momentum. Things moving along a curve will stop only when they lose their angular momentum. To prevent electrons from coming to a halt, Bohr simply decreed that the electron's angular momentum cannot decrease continuously. He took Planck and Einstein's energy element, hv, tossed out the frequency part (because electrons were assumed to be nonvibrating particles) and replaced it with the letter n, which could be any whole number (1, 2, 3, . . . and so on). Then he performed a little mathematics to change h. He gave no physical explanation for his math, but he divided Planck's h by 2π. The quantum h was already extremely small. Dividing it by 2π made it even smaller, but it still was bigger than 0. Multiplying this new small number by n—whose least possible value is one, not zero—ensures that an electron's angular momentum (and, therefore, its energy) cannot decrease forever. Bohr's formula was arbitrary, but it worked.

The fact that Bohr's arbitrary rule worked so well gave him a special reputation among physicists. He did not take facts, explain them, and provide understanding. His genius seemed to rest on intuitions, deep insights based on more than normal powers. Instead of discovering natural explanations as ordinary scientists do (or even imagining the explanations as geniuses do), Bohr intuited successful rules that allowed for the practical continuation of work in physics. Students like Heisenberg were delighted to find a clear pathway to progress in their research, but established physicists who wanted full explanations for their rules were deeply troubled. For them, de Broglie's paper suggested that help was on the way.

De Broglie had proposed that the electron is not a particle orbiting like a planet but a mobile rippling around the atom like a wave. For physicists with a mathematical imagination this idea suggested a meaning for Bohr's whole number, n. Square dancers will have an advantage at visualizing how numbers can determine the size of whirling systems. Imagine four dancers holding hands while swinging around a circle. Now suppose a fifth dancer joins the group. The size of the whirling circle must grow, too, to allow room for the new dancer's arms. Suddenly two of the dancers leave. The remaining dancers must move into a tighter circle so that they can continue to hold hands while moving. De Broglie's orbit is the equivalent of the dancer's whirling circle, and each dancer is the equivalent of one full wave. As the number of waves in an orbit changes, so does it size. Also, just as you need a whole person to be a dancer, so an orbit needs a whole wave. You cannot have partial dancers or partial waves, so Bohr's whole number rule suddenly makes physical sense.

Schrödinger, with his strong background in the physics of waves, immediately recognized a promising mathematical approach to the boundaries forced by whole waves. In this math Schrödinger was luckier than Einstein because, if he was right in his approach, he would not need to learn a new geometry before finding his equation.

After some struggle, Schrödinger derived a draft equation. It seemed to handle all the old cases, so he moved on to the newer problem of quanta and matter. He computed the motion of an electron orbiting a hydrogen atom but the result, regrettably, did not match the experimental data. Schrödinger had to find another approach; however, nothing occurred to him immediately.

Another way that Schrödinger resembled Einstein was in his fondness for and success with women. Possibly he even matched Einstein in the way he grew more passionate as physics grew more frustrating. It is well known to science historians that during this period Schrödinger took a girlfriend to an alpine resort and between—or possibly during—erotic interludes he gained some new ideas. Sadly, we know nothing else about the woman and she must remain the phantom muse of quantum physics.

Schrödinger's new work used an idea studied by his colleague Peter Debye in the years before the war. This was the wave packet, the bulge moving through the bullwhip. Schrödinger calculated that the speed of a moving particle and the speed of a wave packet matched. This kind of identity was the sort of thing that generally brought Einstein up on his toes because it hinted that behind the mathematical match there could be a real, physically meaningful identity. This match between the speed of particle and wave packet, Schrödinger declared, "can be used to establish a much more intimate connection between wave propagation and the motion of the representative point than was ever possible before."

That talk of "the representative point" shows how profoundly Schrödinger's understanding of nature was already changing. "The representative point" was an abstract term for what just a few months earlier he had called corpuscles or particles. Solid matter in his emerging theory was becoming more abstract, a representative of an underlying wave. In this view we can imagine God with a million arms and each of those arms is cracking a bullwhip with a wave packet moving along it. Only, for us mortals in the illusory world of gross sensations, God is invisible, the arms are invisible, and the bullwhips are invisible. All we can see are little points of matter—the merest representatives of the grand system of waves that invisibly comprise the universe's secret reality.

"The true mechanical processes," Schrödinger said, "will be realized and represented appropriately by the *wave processes* . . . and not by the motion of *image points*. . . . Then we *must* proceed from the *wave equation* and not from the fundamental equation of mechanics, in order to include all possible processes."

For the second time in 10 years Newton's physics was being toppled. Oh, sure, as with gravity, the old equations were accurate enough to keep using when solving practical problems, but they could no longer claim to express what was really going on in the universe. Waves, wave packets, and representational points were replacing the forces that pushed the indivisible atomic particles that Newton had believed in.

In the midst of this revolution, Schrödinger's great stride came when he realized he did not have to be quite so revolutionary. He saw that if he altered his equation so that it continued to match Newton's mechanics, but no longer agreed with special relativity, his quantum calculations matched the experimental results. Omitting relativity was no small thing, and Schrödinger, who loved relativity for its power and beauty, knew that a corner of the sheet was staying untucked, but he could see it was progress and took his step.

One appeal of this near miss was the way it promised to get rid of what Schrödinger called "those damn jumps." The electron's habit, in Bohr's account, of jumping from orbit to orbit was too spooky, like the ghost of Hamlet's father—it's here, it's there, it's nowhere in between. Schrödinger's idea built on what he knew about sound waves. Musicians knew about them too. Violinists, for example, can produce a rhythmic beat without the use of percussion instruments, simply by creating two separate notes at the same time. Musical strings can vibrate simultaneously at many different frequencies, giving them the rich harmonic sound that people love. In the hands of a skilled performer these separate notes will produce an audible beat, the vibrato that Einstein's own playing so notoriously lacked. A sensitive vibrato throbs like a heart rather than ticks like a clock. The sounds are produced by shifting frequencies. The sound reflects the differences between the wavelengths of the various, simultaneous notes, and the throbbing comes from the way the violinist keeps changing the string lengths so that the notes keep changing ever so slightly.

Schrödinger proposed that the changes in electron emissions could be understood as a kind of atomic vibrato. As the frequencies of an electron shifted slightly, they would create slightly different harmonic beats that would look like fluctuating pulses of energy. Schrödinger commented, "It is hardly necessary to point out how much more gratifying it would be to conceive of the quantum transition as an energy change from one vibrational mode to another than to regard it as a jumping of electrons."

Exciting as all this was, Schrödinger found that the more deeply he probed, the less physically real his waves became. His theory appeared in a series of four articles published between January and June, 1926. In each paper, the wave became a bit less part of nature and a shade more a part of mathematics. The theory's basic problem was that it did not make good physical sense. What are these waves? Yes, the speed of a particle matches the speed of a wave packet, but where do all the elements of the basic wave form enter the story? The wave form is familiar to anybody who ever stood on a pier and watched the sea lapping below. Waves vary in height and length. Schrödinger's problem was that none of these elements appear in his final equation for the particle-packet's velocity. What kind of wave can there be when it has none of the wave form?

"If," said Schrödinger in his final paper, "we regard the whole analogy [between optical waves and mechanical movement] merely as a convenient means of picturization, then the defect is not disturbing, and we would consider any attempt at improving upon it as idle trifling." That view might satisfy some, but Einstein—and Schrödinger too, on days when he was not in love with an idea—always wanted to know what was *really* going on.

And Schrödinger saw a second problem if his system was taken literally. "Classical mechanics," he maintained, "breaks down for the very small dimensions and very great curvatures of path." There can hardly be a smaller dimension and tighter curve than those encountered by an electron buzzing around an atomic nucleus.

So Schrödinger no longer felt confident that his waves were real and he deliberately chose an abstract symbol—ψ, called psi, or p-sigh—to represent the solution to his equation. "One may, of course," he commented, "be tempted to associate the function psi with a vibrational process in the atom, a process possibly more real than electronic orbits whose reality is being very much questioned nowadays. Originally, I, too, intended to lay the foundations for the new formulation of the quantum conditions in this more intuitive manner, but later I preferred to present them in . . . neutral mathematical form."

For Schrödinger, the strength of this approach remained what had been its most compelling attraction when he first read de Broglie. It got rid of Bohr's arbitrary rules for inserting whole number ns into quantum equations and replaced them with a coherent measure of waves occupying space. Yet even that space was no longer the kind of everyday space familiar to anybody who ever got lost in a woods. It was an abstract, mathematical "space" with its own rules and properties. As Schrödinger put it in his fourth and final paper on undulatory mechanics, "The psi function is to do no more and no less than to offer us a survey and mastery over [electrodynamic] fluctuations by [using] a single differential equation. It has been repeatedly pointed out that the psi function itself cannot and may not in general be interpreted directly in terms of three-dimensional space . . . because it is in general a function in [a mathematical] space and not in real space."

Nevertheless, Schrödinger's close attention to what was really going on was much appreciated by Einstein and his colleagues. Planck responded early to the articles, writing Schrödinger, "You can imagine the interest and enthusiasm with which I plunge into this study of these epoch-making works."

Lorentz wrote, "If one can successfully explain the phenomenon [of fluctuating electron radiation] by connecting a definite frequency to the moving electrons, it would be much more beautiful" than in rival, more arbitrary theories.

Einstein wrote him too, "Professor Planck pointed your theory out to me with well justified enthusiasm, and then I studied it with the greatest enthusiasm. . . . The idea of your article shows real genius."

This note delighted Schrödinger who promptly replied, "Your approval and Planck's mean more to me than that of half the world. Besides the whole thing would certainly not have originated yet, and perhaps never would have (I mean, not from me), if I had not had the importance of de Broglie's ideas really brought home to me by your second paper on gas degeneracy."

Part III

A Radical
Understanding Defied

23

Sorcerer's Multiplication

"A new reality is in the air," began the title song of one of Berlin's hit revues of 1926. The idea of a new reality, the *Neue Sachlichkeit*, was part of the mid-1920's mood that shaped Berliners. The war had smashed the old social order. The inflation had destroyed middle-class morals based on saving for a rainy day. Change was everywhere and Berliners believed in progress. They had not been this confident and optimistic since the war's early days. Construction work was peppered across the city. Radio broadcasts sounded from apartment windows. Women bobbed their hair and, in 1926, the rising hemline reached the knee. The style was, to use a term favored by the Germans, "American"—meaning the style was up-to-date, rational, functional, secular, successful.

Hindsight knows this optimism was another lie. A different modernity was soon to come, but the new Weimar reality might not have been an impossible dream. It was open, complex, human and, therefore, sorely resented by the millions who had loved the old lies and missed them still. Who would be quicker on their feet, the republic's beneficiaries or its enemies? The enemies were organizing themselves just offstage. Hitler had tried to blast his way into the spotlight and failed, but he had not given up hope, and he fought to strengthen his grip over the Nazi movement. His rivals in the party were supported by an energetic and adept young rabble-rouser named Joseph Goebbels. In early 1926, Hitler won Goebbels to his side. The new loyalist would prove himself a liar *extraordinaire* and would begin feeding Germany's old addiction to falsehoods.

But we are getting ahead of ourselves. At the time, internal shifts in crackpot organizations were too common and trivial to notice. The more promising struggle, the one that held Einstein's full attention, was a search for the truth. Einstein in this new Berlin was both modern and old-fashioned. He participated in the changes as a man-about-town. He went to the shows, met the stylish people, and discussed the newest ideas. Yet he was also an old-style savant, a thinker who had not become a journalist, a visionary who did not turn to photography, a musician who did not enjoy the new dissonance. That old-fashioned seeker after truth was the enduring part; Einstein's modernity was merely part of his Berlin wardrobe, and his clothes were never that important to him.

A few days after receiving Schrödinger's letter, Einstein sent him a second note, "I am convinced that you have made a decisive advance with your formulation of the quantum condition, just as I am equally convinced that the Heisenberg-Born route is off the track." That reference to a rival approach reflected the peculiar state of quantum physics in the spring of 1926—after years of bafflement, physicists all over Europe were having their bathtub moments and shouting "Eureka." Heisenberg had developed a "quantum mechanics" at Göttingen, which his advisor, Max Born, had given a more coherent mathematical form. A graduate student in Cambridge, Paul Dirac, had proposed a mathematical theory of quantum behavior that has become known as q-number algebra. A few months later, Erwin Schrödinger announced from Zürich his new undulatory theory. The quantum revolution that Einstein had been anticipating for a quarter of a century had begun at last, and, as in any revolution, past experience provided only the crudest guide to its course.

Science was well acquainted with the phenomenon of separate discoveries of the same thing. Darwin was shocked when he read Alfred Wallace's paper and found that his own chapter titles appeared as phrases in Wallace's theory. But the quantum eurekas of 1925 and

1926 appeared, in a startling way, mutually contradictory. All these new ideas were being born, all of them appeared to work—in the sense that they correctly predicted changes in quantum states—and each of them meant something profoundly different. There had not been such confusion in physics since Galileo's time, when three rival theories of the universe stood side by side. The available data supported each of the theories. Choosing one depended on instinct. Intelligence alone could not point the way.

By the spring of 1926, when Einstein congratulated Schrödinger for making "a decisive advance," the role of instinct had returned to science. By praising Schrödinger, Einstein meant what he always meant when he spoke of a step or a stride forward: theory had combined with meaning to give a new ability to understand phenomena. When Einstein said others had gone "off the track," he was objecting to the absence of meaning from their theories.

Even the staunchest defenders of Heisenberg's and Dirac's new ideas did not pretend that they had yet clarified underlying meaning. They worked in the tradition of Bohr's atom, replacing explanation with mathematical decrees. Dirac's q-number algebra, for example, was almost completely abstract. Dirac had devised a new type of number with all manner of unusual properties. Number a, for instance, was different from number b, but neither one was bigger than the other. How then, one might ask, did they differ? Whatever the answer, it was quite divorced from the familiar experience of counting or measuring. Dirac provided rules for manipulating his numbers, and when physicists followed the rules they got results that matched experiment, but as for understanding what it all meant, the physicists might just as well have been consulting the hieroglyphs of a lost language. Meanwhile, the quantum mechanics of Heisenberg and Born that Einstein had dismissed in his note to Schrödinger was hardly less abstract. Its most remarkable feature was that the order of calculation suddenly mattered. If, so to speak, you multiplied 2 times 3, you got a different answer than if you had multiplied 3 times 2. "A real sorcerer's multiplication table," Einstein laughed.

Heisenberg prefaced his system with the remark, "The present paper seeks to establish a basis for theoretical quantum mechanics founded exclusively upon relationships between quantities which in principle are observable." Focusing "exclusively" on "quantities" meant that Heisenberg had taken the astonishingly bold step of removing geometry from his physics. His method took a quantum state such as its energy level, performed a mathematical manipulation, and reported a new quantum state. How the particle got from one state to the other was unspecified.

Einstein was immediately offended by this notion and demanded of Heisenberg, "You don't seriously believe that none but observable magnitudes must go into a physical theory?"

The challenge seems to have caught Heisenberg by surprise. As he recalled it many years later, he replied that Einstein had done exactly the same thing in his theory of relativity.

Einstein might have shrieked, "I did not." Neither the relativity principle nor the constant speed of light, his famous theory's axioms, were observable magnitudes. However, as Heisenberg recalled it, Einstein muttered something about, "A good trick should not be tried twice," but then he added, "In reality the very opposite happens; it is theory that decides what we can observe." The retort appears to have silenced Heisenberg, at least for the moment. "Measuring observables" sounds like it is free of theory, but theory defines the measurement. For centuries astronomers had measured Mars's position in the sky and believed they were recording the absolute motion of Mars around the earth. Then Kepler made measurements, believing them to trace Mars's orbit around the sun, relative to the earth, and a new physics of the sky was born. It was this long experience that had led Einstein to tell the Prussian Academy during his inaugural address that, without principles, "the individual empirical fact is of no use to the theorist; indeed he cannot do anything with isolated general laws abstracted from experience."

This new quantum mechanics had lain in the background of the November 1925 meeting between Bose and Einstein. When Einstein had asked Bose if he could press his statistical approach further to present a new explanation of quantum change, he was asking if Bose could connect a new quantum theory with the old quantum facts.

At the Lorentz celebration in December, quantum mechanics had been the great new idea behind many discussions. Every month the revolution brought new techniques for manipulating quantum data. November produced the "three-man paper" of Max Born, Werner Heisenberg, and Pasqual Jordan. It constructed a firm mathematical foundation for Heisenberg's approach, but this foundation rested on a branch of mathematics called matrix calculus that was new to physics. Most physicists had no idea how it worked, and the joke going around the labs that season was that physics was getting too difficult for the physicists. So in early 1926, when Schrödinger's wave mechanics offered a more familiar mathematics and its meaning that particles are, at root, wave packets, there was a sigh of relief from physicists who liked their science meaningful and doable.

Five years earlier, instinct had held a much more subtle role in physics. It made some people experimentalists and some theorists. It pointed some quantum investigators toward radiation and some toward matter, but authority had always come from the facts. The fact of the Compton effect and follow-up experiments had settled the issue of light quanta, no matter what a person might have preferred. The Bose-Einstein statistics had met little resistance because they were based on those photons that were fully supported by the experimental data. The facts about quanta and matter, however, were different. They pointed nowhere—you had one quantum state, then another. Without a theory to explain those facts, scientists were free to use whichever technique they preferred for calculating the changes. As most people are pragmatists about their methods and anti-revolutionary by temperament, the mass of physicists preferred to use Schrödinger's equation. Its form was more familiar than matrices or q-numbers, and it was also easier to use. Matrix methods challenged the intellectual habits of a lifetime by abolishing any role for geometry from its calculations. Yes, Einstein had already demanded extensive rethinking of space and time, and he had gotten it, but Schrödinger's wave equation made that kind of radicalism unnecessary.

Still, there were physicists whose motives were more personal than practical or professional. Einstein was never one to favor the easier or more familiar pathway. Neither was Max Born nor Niels Bohr. Those who joined Einstein in preferring Schrödinger for other than practical reasons usually agreed that they wanted to understand the fundamentals and to know what sat at the base of reality. For them, facts by themselves were trivial, interesting only for their ability to point at the deeper truth of what lies at the bottom of natural events. It was that ambition that led Planck to speak enthusiastically to Einstein about the Schrödinger papers. Wilhelm Wien made himself the most public spokesman for this attitude.

These battles became so bitter that each group characterized the other so crudely they might have been hack political propagandists. Even today it is too easy to find accounts of the quantum revolution that portray it as a battle between conservatives and progressives, or old versus young, or of students of Kant against modern positivists. None of that rhetoric helps us much in understanding the behavior of particular scientists. It seems absurd to speak of skeptical Einstein as a conservative, or his contemporary Max Born as a youth, or of philosophically indifferent Heisenberg as anticipating the positivism of the Vienna Circle. The split was much more sharply focused on the nature of physics. The Einstein side was the heir to the Euclid–Newton tradition that systematized facts and gave them meaning for the nontechnical world.

The unEinsteinian school was more technically oriented and was more dubious about finding general truth. Their attitude had been suppressed during Newton's long glory, but for most of the history of science, theirs was the dominant notion. They wanted rules that worked and let them do their physics; they did not want to bother with the larger meaning of it all.

This difference explains why—with the exception of his discovery of the photon—Einstein always denied that his work was revolutionary. True, he had demoted both Newton and Euclid from the ranks of divinely inspired prophets, but he had also kept to their goals of understanding the meaning of laws. When he toppled Euclid, Einstein immediately set an alternate geometry in its place. Surely he never expected to read, as one could find in the three-man-paper, "Admittedly . . . a system of quantum-theoretical relations between observable quantities . . . [labors] under the disadvantage of not being . . . geometrically visualizable since the motion of electrons cannot be described in terms of the basic concepts of space and time." For Einstein and the other opponents of quantum mechanics, this was no step forward. Embracing it meant throwing out rational, accessible nature.

During the battle over the Compton effect, Wien had published his newspaper article that insisted no true physicist would ever accept the BKS theory of random, arbitrary motions. Later, in November, 1925, at about the same time that Bose and Einstein met, Wien gave a speech in Munich that proclaimed how Newton's laws "had revealed to man for the first time the possibility of comprehending nature by the logical force of his intellect." The following June, at Munich University's Founder's Day celebrations, Wien insisted that no part of physics is barred from human understanding. Physicists, he insisted, would not rest until they understood atomic processes. Rejecting mathematical manipulation, he promised that "number mysticism would be supplanted by the cool logic of physical thought."

It was in that confident expectation of being on the road to finding the universe's rational basis that Einstein wrote his old chum from college and patent-office days, Michael Besso, bringing him up to date and telling him, "Schrödinger has come out with a pair of wonderful papers on the quantum rules." That letter was written on May Day, 1926. It was the last time Einstein would make an unambiguously enthusiastic remark about the progress of the quantum revolution.

24

Adding Two Nonsenses

There is something of a joke in the way that, just as physicists began warring over the truth of physical law, the rest of the world's avant-garde was insisting that their subjects were lawful, too. Nonphysicists have, for centuries, looked jealously at the order and solidity of physical foundations and have lusted after it for their own fields. Lavoisier's physics-envy brought reasoned order to chemistry. Lyell's imported it into geology. There were also efforts to bring systematic law and order into the study of human affairs—Karl Marx in history, John Watson in psychology, and heaven knows how many thinkers in economics. Most of these efforts to link people and natural law were failures. And when it was tried in politics, as with bolshevism and Nazism, it became literally criminal. In the arts, however, particularly in literature, the fruit was sometimes more satisfying.

Proust's first volume tells of a certain Monsieur Swann who traveled in France's most elegant circles, dining at the Jockey Club, playing cards with the Prince of Wales, and yet never speaking of these connections. His neighbors in the small town of Combray have no idea of the distinguished company he keeps. In volume 2, readers immediately learn that Swann is to be avoided because he is always bragging about the nobodies who have entertained him and his wife. No, Swann's character has not changed. It turns out that Swann has exactly the same character he had before, but he lives in new circumstances and is seen by himself and by others in a new way. Poor Proust, he hoped that there was some relation between his account of time and Einstein's. There was none. He died without knowing that when it came to a belief in a natural law that underlay seemingly contradictory events, he and Einstein were one.

But there was another move afoot that was equally modern and, because it was more practical, had already seized hold on the lives of millions. This was the effort to give technical rationality to the operation of social systems. The moderns behind this transformation based their laws on efficiency rather than meaning; naturally, that kept their focus narrow. Yet it let them go deep into the details, so they could scorn the broad-minded generalists who lacked their expertise.

Straddling these two modernisms was Niels Bohr. He was born in the late nineteenth century, when meaningful systems appeared triumphant: Species had been linked through an evolutionary history, chemical elements through a periodic chart, and even the impulses of the mind through a psychoanalytic theory. Bohr had no instinct for such systematic thinking, as was probably best shown in his distaste for mathematics. It seems hard to believe that such an anti-arithmetician could have risen so high in theoretical physics, and it often startled the wonderlads of physics when they first met him. Paul Dirac heard Bohr when he spoke in Cambridge in May, 1925. "People were pretty well spellbound by what Bohr said," he remembered, "While I was very much impressed . . . his arguments were mainly of a qualitative nature and I was not able really to pinpoint the facts behind them. What I wanted was statements which could be expressed in terms of equations, and Bohr's work very seldom provided such statements. . . . He certainly did not have a direct influence because he did not stimulate one to think of new equations."

If Bohr was ever to sympathize with quantitative arguments, the new mathematics of Heisenberg, Born, and Dirac provided the sorts of equations that best suited him. They allowed great feats without pointing toward deeper physical truths, and Bohr was enthusiastic about this practical math from the start. He first learned of Heisenberg's quantum mechanics from a letter in which Heisenberg made a brief, happy reference to his new paper. At a mathematics congress, Bohr then reported that Heisenberg had made a breakthrough, "probably of extraordinary scope." Bohr's immediate embrace of quantum mechanics came on the heels of the collapse of the Copenhagen achievement in quantum theory. The success of the Compton effect and the defeat of the BKS strategy had been a great blow. Similarly, his 1913 model of the atom had become increasingly hard to credit, and his protégés were saying to his face that they doubted its key feature—electrons following fixed orbits around an atomic nucleus.

"The most important question seems to me," Pauli wrote Bohr, "to be this: *to what extent may definite orbits of electrons in the stationary states be spoken of at all.* I believe this can in no way be assumed as self-evident." By getting rid of geometry, the new quantum mechanics abolished Bohr's famous electron orbits and jumps just as decisively as Schrödinger's wave mechanics would do. Yet Bohr embraced it all, and without a fight.

Not everybody followed his example. In November, 1925, Heisenberg told Pauli that Göttingen physicists had already fallen into two camps, those who welcomed the success of matrix algebra, regardless of its abstractions, and those who denied it was even physics. Bohr was with Heisenberg because he liked unambiguous abstractions. Physicists were always treating their abstractions as real—like energy, so real it could be conserved. But with matrices there was no danger of that temptation toward realism.

Bohr's anti-realism prejudices are often attributed to one philosophy or another, but that is the suspicion of people with very few such tendencies of their own. It is like thinking a religious person must have fallen under the sway of some preacher, and disregarding the possibility of private experience and personal taste. Einstein was the one who read philosophers for his own entertainment. Years later, when the heroic age of quantum physics was well in the past, Bohr did attend a conference of philosophers. He reported afterward, "I have made a great discovery, a very great discovery. All that philosophers have ever written is pure drivel."

Instead of the organizing logic that Einstein hoped for, Bohr wanted techniques that would allow him to consider particular physical problems. His great technique had been the correspondence principle, the rule of thumb that allowed him to approximate quantum results by thinking in classical terms. He was not interested in resolving the contradictions between quantum and classical physics so much as he was in using them. And his students, all of whom were more mathematically oriented than their teacher, absorbed this attitude.

Heisenberg once said, "I learned optimism from Sommerfeld, mathematics in Göttingen, and physics from Bohr." And the physics that Bohr taught was the unambiguous, unrealistic variety. Theoretical contradictions—Einstein's key to progress—were not a worry to Bohr. He later recalled that in those days of struggle, when nothing worked, his students at the Institute consoled themselves with the joke that along with "statements so simple and clear that the opposite assertion obviously could not be defended" there were also "'deep truths' in which the opposite also contains deep truth."

Bohr looked for this attitude in students. He finally invited Pauli to come to Copenhagen when Pauli wrote that quantum physics was mired in two great bits of nonsense and that "the physicist who finally succeeds in adding these two nonsenses will gain the truth!"

The taste for nonsense probably lay behind the skepticism that Bohr displayed about electron spin when he arrived to celebrate Lorentz's 50 years as a doctor of philosophy. Spin had first been proposed in Copenhagen by a German-American physicist named Ralph Kronig. Heisenberg and Pauli immediately hated the idea. They had just chased all imaginable actions out of quantum mechanics. Now Kronig was proposing to set the electron rotating in space. Faced with such fierce criticism by Bohr's stars, Kronig grew silent and the idea of electron spin had to wait for a second coming, this time under Ehrenfest's encouraging eye.

That was the state of physics during this period. With new ideas coming as regularly as the full moon, nobody's method could absorb them all. It was a predicament that suited Bohr's practicality. Heisenberg's method was to stick with his matrix calculations, confident that eventually it must lead to "all the right answers." Schrödinger held the same attitude toward his wave mechanics. But Bohr's attitude, as summed up by Heisenberg was, "Well, there's one mathematical tool—that's matrix mechanics. There's another one, that's wave mechanics. And there may still be other ones." Bohr was entirely flexible in this matter because, unlike every other physicist struggling to understand quantum changes, his ultimate interest lay in finding an unambiguous, qualitative physics.

Bohr was explicit about his preference for nonquantitative physics. When he was finally persuaded to publicly endorse the concept of electron spin, he wrote that spin "opens up a very hopeful prospect of our being able to account more extensively for the properties of elements by means of mechanical models, at least in a qualitative way characteristic of applications of the correspondence principle." As so often the case with Bohr's remarks, this passage's meaning seems to dissolve when examined closely, but we can get its sprit. Spin offers hope of more qualitative understanding, and more support for the beloved "magic wand," the correspondence principle.

Mathematically oriented physicists were perhaps surprised by Bohr's remark, because Schrödinger's equation had given quantitative rigor to the correspondence principle and thereby done away with the need for a rule of thumb. In the spring of 1926, the rising popularity and strength of Schrödinger's equation looked to mark the ceiling for Bohr's fame and achievement. The old quantum theory, in which he had been a founding revolutionary, was being replaced by a new mechanics that digested the best of his old ideas, giving them a rigor they never had before. And then in early May came the surprise that ultimately set Einstein to defiance and raised Bohr to a sage.

25

Admiration and Suspicion

E instein always loved his violin and played it wherever he traveled. Musicians were welcomed into the Einstein apartment to play. He sometimes performed in public as well. Einstein occasionally played during dusk services at Berlin's New Synagogue. There was room for 3,000 people. The men filled the main floor; the women sat silently in the balcony while the organist played Bach. Einstein accompanied on his violin and the city's fading daylight crept through stained-glass windows. Those sounds in the darkness were Einstein's escape. "Whenever he felt that he had come to the end of the road . . .," his son Hans Albert recalled, "he would take refuge in music and that would usually resolve all his difficulties." It is too happy an interpretation. In the morning, the difficulties, especially the meaning of quanta, persisted. They could rebound with unanticipated strength.

The surprise challenge of May 1926 was the publication of Schrödinger's proof that wave mechanics and matrix algebra were mathematically equivalent. This idea was much in the air. Schrödinger had submitted his article in mid-March. At March's end, an American named Carl Eckart sent a paper to the National Academy of Sciences making the same argument, and in April, Wolfgang Pauli outlined a proof in a private letter to Born's assistant, Pasqual Jordan.

Schrödinger's proof took the most peculiar feature of matrix algebra—the fact that p times q does not equal q times p—and showed that he found the same peculiarity when using his wave equation's differential calculus. Schrödinger then showed that any matrix equation could be translated into his wave mechanics and, conversely, any wave function could be translated into matrix math.

This formal unity served the function of a scandalous moment at a dinner party. It revealed something serious and different about each person who was witness to the incident. Heisenberg, Schrödinger, Bohr, Einstein, and Born each responded separately. The immediate question was what this mathematical equivalence meant. Were the equations physically equivalent too? Heisenberg became much more ardent in insisting that his system was the correct one, and that Schrödinger's wave mechanics was "disgusting." Schrödinger and some other wave enthusiasts were equally determined to defend their system.

For Einstein, the situation showed that the revolution still had a long way to go. The mathematical equivalence reflected the fact that neither one got down to nature's deeper truth.

Max Born's reaction was the most distinctive. He usually focused on keeping the math rigorous and clear, a taste that made him seem a bit odd to many other physicists. Even his assistant, Wolfgang Pauli, once complained to Born about his "tedious and complicated formalisms" and how he spoiled good physical ideas by his "futile mathematics." Presumably Pauli meant that physical meaning was not the same thing as mathematical meaning, yet Born looked to physics' mathematical side. Born's approach, however, became a great strength during the birth of quantum mechanics. Although nobody was having any luck finding physical meanings for the discoveries, mathematical meanings were there for the taking. A grand example of that success lay in the birth of quantum mechanics. Heisenberg had been looking for a math system that could predict changes in a particle's quantum state. He finally succeeded in drafting a paper and giving it to Born to look at. Heisenberg then immediately left Göttingen for vacation. When Born read the paper, he saw the peculiarity that p times q did not equal q times p and he knew that this oddity must mean something important.

Most physicists presented with such an oddity would look for some insight into physical reality. What property might nature have that would give such a result? But Born looked for a mathematical meaning. He remembered that he had already encountered the same thing in his student days. So for him, Heisenberg's paper meant that quantum mechanics rests on matrix algebra.

Einstein would have pressed on to ask what was found in nature that required matrix algebra; however, for Born, matrix algebra was the whole solution. He translated Heisenberg's confusing notation into matrix form, and was well pleased because the mathematics now made sense. In matrix algebra, the multipliers (p and q) are not numbers but systems of numbers and multiplication is a method of combining number systems.

Suppose we have two sets of letters, IE and TM, which we can symbolize by p and q. We can combine the sets by alternating letters from each group. We take the first letter from one group and then the first letter from the next group; next we take the second letter from the first group and then the second letter from the second group. If we begin with the p group, pq = ITEM, but if we begin with the q group, qp = TIME. There is nothing inherently puzzling in this case about pq giving us a different result from qp, but while this mathematical understanding might relieve the feeling that the situation is impossible, it does nothing to explain the reality of what lies behind quantum mechanics. If you take the matrix algebra to be the whole story, normal physical explanations go out the window. There is only a mathematical explanation, just as there is only a mathematical reason for why three oranges priced at 30 cents apiece cost 90 cents. There is no natural cause at work here, nothing about the properties of oranges or pennies. There is just the mathematical fact that 3 times 30 equals 90. Similarly, the changes in quantum states have no underlying physical explanation. They are what they are because of the mathematics of how matrices work.

It was this kind of abracadabra that led Einstein in the summer of 1926 to write Ehrenfest, "I look upon quantum mechanics with admiration and suspicion." Who could not help but admire the multiplication table that allows us to find a price for any number of oranges?

Yet how can you not be suspicious that this system misses much about an orange's quality, size, and freshness?

Schrödinger's proposal to explain quantum states in terms of waves and wave packets had suggested a method that could look beyond the mysterious matrices to reveal what was really going on. But in June 1926, Max Born submitted a paper that reinterpreted Schrödinger's equation, using it to consider what happens when atomic particles collide. The matrix approach to this issue was to define a quantum state at the instant of collision and a second state after the collision, with no causal process between the input and the outcome. Schrödinger's method viewed the collision as the interaction of two wave packets and calculated their behavior through space and time. In this view, every part of the process from start to finish is connected. But Max Born proposed a new way of thinking, one he credited to Einstein's influence.

Born was unwilling to abandon material reality quite as thoroughly as Heisenberg did and he appreciated the way Schrödinger's method kept space in physics' story. Yet Born refused to believe that particles were illusions, wave packets that passed themselves off as something more. Still, Schrödinger's equation described something; if it did not describe a wave, what was it about? Born was back in the same blind alley that quantum work kept entering. How do you reconcile waves and particles? But Max Born's instinct was to look for a mathematical reconciliation rather than a physical one.

One possibility was that instead of defining a wave, Schrödinger's equation defined a field. After all, electromagnetic fields played a large role in the behavior of quantum radiation and particles. Born recalled the "ghost field" idea that had bothered Einstein for so many years. The same old difficulty showed itself. What was this field? How did it work?

Faced with the question that never seemed to go away, Max Born took what many physicists consider one of the grandest strides in the history of science. At the time, however, the step was not universally appreciated and Einstein always considered it a terrible move backward. Born stripped the field of any physical reality whatsoever and gave it a purely mathematical meaning. The wave or field in

Schrödinger's equation—the thing that was represented by ψ—was a statement of probability. As Born put it, "One obtains the answer to the question, not 'what is the state after the collision,' but 'how probable is a specified outcome of the collisions.'"

The radicality of Born's idea was that there was no physical explanation anywhere behind these probabilities. It was not like a weather forecast in which the chances refer to physical events. If there is a 50-50 chance of rain, the forecaster is saying, in effect, "Well, the clouds might go this way or they might go that way." Whatever way they do move is determined by meteorological forces too subtle for the forecaster to anticipate, but once it is all done the forecaster can say, "We didn't get that rain after all because that high pressure system came in and pushed the rain clouds away." The forecast was given in terms of probabilities, but the events were all physical.

Max Born's probability field had no such physical meaning. Born then made an even more radical assertion, saying that in "quantum mechanics there exists no quantity which in an individual case causally determines the effect of a collision. . . . I myself tend to give up determinism in the atomic world."

Born was taking the position Einstein had always refused to take. Einstein had added two extra years to his struggle with general relativity because he insisted on finding a causal explanation for his equation, and, after general relativity, he again had a chance to abandon causality when he published a theory of spontaneous emission of light quanta. The trouble with the theory, as Einstein noted at the time, was that it could not predict where the light would go. Ever since then, Einstein kept an eye open for some new idea that would complete his theory by predicting the light's direction. Now Born was saying there was no need to spend years looking for an explanation. At the quantum level, things do as they damn well please.

This was the return of BKS with a vengeance. Niels Bohr had proposed that quanta behave statistically, not on the basis of what had come before. Einstein had said then that he would prefer to work in a casino than to be a physicist in such a world. BKS had failed, but Max Born had restored its croupier spirit. Or almost restored it. The triumph of Einstein's photons in the Compton effect had vanquished the notion of an absolutely statistical universe in which nothing had to be conserved. Particles, in Born's account, moved statistically, just as they did in the BKS paper, but the probabilities of those motions were absolutely fixed in accordance with Schrödinger's wave field.

Viewed from that perspective, Max Born's paper on particle collisions marked the conservative cooling point, the Thermidor, of the quantum revolution. The most radical elements—Niels Bohr's assault on conservation and Heisenberg's attempt to abolish space from quantum considerations—failed. But Born was no counterrevolutionary and he did not restore the old regime. Statistics, rather than reasons, governed the outcomes of collisions; space was filled with probabilities rather than motion. The particle was there; now it is here. As the philosopher David Albert put it much later, "Electrons [in an experiment] do not take [one] route and do not take [the other] route and do not take both of those routes and do not take neither of those routes." Baffling, wouldn't you say?

If we return for a moment to the metaphor of the bullwhip, in Max Born's interpretation the whip is gone and the wave packet is gone, but sometimes the cigarette still flies out of the lovely assistant's lips.

Whenever a revolution cools there are people who resist its halt. Like America's whiskey rebels after the establishment of the federal constitution, they grumble that this outcome was not what they were looking for when they enlisted in the sacred cause. Rarely, however, does a living symbol of the revolution resist its settlement quite so ferociously and famously as Einstein resisted this one. He did see much to admire. The new quantum mechanics was a technical achievement of enormous imagination. Whatever finally emerged would have to be just as accurate as the existing mathematics. Quite possibly, the existing mathematics would survive but be understood in a new way. This kind of transformation had happened before. Einstein's theory of relativity had found a new way to derive and explain Lorentz's equations. Bose's paper had provided a new way to derive and explain Planck's Law. Einstein was confident that, now that it was making progress, the quantum revolution had to heat up still more. After decades of struggle, physicists had a new toolkit to work with, but Einstein never thought tools alone led anywhere. Technique by itself is not enough to advance a civilization from Babylonian practicality to Greek understanding.

In the summer of 1926, it was still not clear where the quantum revolution would end. Colleagues assumed they would remain colleagues. Max Born did not expect to break with Einstein, and in his second paper on wave mechanics he publicized his debt to his great friend, "I start from a remark by Einstein on the relation between wave field and light quanta. He said approximately that the waves are only there to show the way to the corpuscular light-quanta, and talked in this sense of a 'ghost field' that determines the probability for a light-quantum . . . to take a different path." Born's recollection puts Einstein suspiciously in tune with the probabilistic interpretation, but the gist is plain. Einstein had long toyed with the idea, yet he never published it. In his 1917 paper on light quanta, Einstein had objected that his work did not get physics "any closer to making the connection with wave theory." He could—and did—have the same doubts about probabilistic waves. It gets us no closer to understanding how and why particles move through space. As he later wrote, he feared that the emerging quantum mechanics offered "no useful departure for future development." If quantum mechanics was a step, Einstein feared it was a stride into a box canyon and could never lead to an understanding of what lay at the heart of the universe.

26

An Unrelenting Fanatic

I n October 1926, Schrödinger arrived by train in Copenhagen. Niels Bohr was at the station to welcome him and the two began talking immediately as they made their way to the tram. Bohr had invited Schrödinger up north to give a lecture on the "Foundation of Undulatory Mechanics," although Bohr did not believe a word of the theory. Proving the mathematical equivalence of all existing forms of quantum mechanics had merely intensified the ferocity of the dispute between the partisans on either side. Quantum physicists had become like philosophers who agree on every point about, say, archaeology, excepting only whether it should be called a science or not. With so much authority resting on so slender a hook the disputants showed no mercy.

Heisenberg had become Bohr's chief assistant and he witnessed the quarrel. It went on for days while Schrödinger was a guest in Bohr's house. "Although Bohr was otherwise most considerate and amiable in his dealings with people," Heisenberg recalled, "he now appeared to me almost as an unrelenting fanatic." Schrödinger had begun to face the full Bohr treatment, the sort that had eventually turned Kramers into Bohr's lapdog.

Bohr had, from the beginning of his days as Rutherford's student, come at quanta from the material side of things. He had been the one to link quanta and atoms. Schrödinger came at quanta from another point, arriving via de Broglie, Bose, Einstein, and ultimately Planck and his little chunk of radiation, hv. Einstein's $E = mc^2$ meant that ultimately atoms and radiation were like matrix algebra and wave mechanics—different forms of the same thing. But they did not feel like the same thing and the people who came at quanta from these separate directions had different tastes.

Bohr liked the jumps of quantum mechanics and their defiance of geometry.

"If all this damned quantum jumping were really here to stay," Schrödinger gave voice to his exasperation, "then I should be sorry I ever got involved with quantum theory."

Bohr, Heisenberg said, "was not prepared to make a single concession to his discussion partner."

Schrödinger held his ground. Wave theory had formidable strengths. In his mechanics, matter and energy passed through space in the usual way of moving continuously from point to point and without ever disappearing down a rabbit hole.

Heisenberg appears to have prudently stood aside during this dispute, although wave theory's lack of physical surprises was the very thing he disliked most. The world of quanta seemed too weird to be accounted for in the ordinary way of space, time, and motion.

Bohr and Schrödinger continued disputing relentlessly, like nineteenth-century boxers before there were limits set to the number of rounds. Neither had a knockout punch left in him, but neither was willing to throw in a towel.

Bohr rather liked Max Born's suggestion of a probabilistic interpretation of the psi wave.

Schrödinger would have none of this rejection of physical explanation.

Bohr thought particles could not be described as wave packets because wave packets become unstable over time.

Schrödinger thought that matter jumping from point to point—"liberated" from the constraints of space and time—was absurd.

"It will hardly be possible," Heisenberg recalled, "to convey the intensity of the passion with which the discussions were conducted on both sides."

Bohr would not, according to Heisenberg, "tolerate the slightest obscurity," on Schrödinger's part, but, of course, as usual, Bohr himself was deeply obscure.

Schrödinger said Bohr talked "often for minutes almost in a dreamlike, visionary and really quite unclear manner, partly because he is so full of consideration and constantly hesitates—fearing that the other might take a statement of his point of view as an insufficient appreciation of the other's."

Bohr argued like a thoughtful dentist who is persistent in his drilling but regularly interrupts his labor to ask, "Does it hurt? Let me know if it hurts."

After days of this treatment, perhaps by coincidence, perhaps not, Schrödinger became ill and had to stay in bed. Mrs. Bohr proved as faithful a nurse as Walt Whitman, attending the guest regularly, and spoon-feeding him broth. All the while, Bohr sat nearby speaking of the folly of taking wave mechanics literally as a picture of what really goes on in the world.

And when it was time for Schrödinger to leave Copenhagen, neither side had conceded any ground. Neither side could answer all the objections of the other, and neither could give a complete account of how his favored system was supposed to work. That was the way of it during quantum's heroic era: the toolkit grew considerably while the meaning of it all was as sharply disputed as the meaning of a close election.

27

The Secret of the Old One

Goebbels arrived in Berlin that November. He liked to recall it as a grand moment, something like Lenin's arrival at the Finland Station with party members on hand to escort him before an eager audience. In reality, however, the Berlin Nazis were almost nonexistent. That was why Hitler had sent him there, to begin sowing and cultivating his lies as the new *Gauleiter* (director) of Berlin.

Elsewhere, optimists were feeling like they had found some truth. That same November, Max Born wrote to Einstein about his continuing pleasure with his probabilistic interpretation: "About me it can be told that physicswise I am entirely satisfied since my idea to look upon Schrödinger's wave field as a 'ghost field' in your sense...."

Presumably Einstein winced at that "in your sense." It was like Marx thanking Hegel for dialectical logic.

"Schrödinger's achievement," Born continued, "reduces itself to something purely mathematical; his physics is quite wretched."

Born had every reason that autumn to be pleased with himself. He had cleaned up both main versions of quantum mechanics—putting Heisenberg on a clear mathematical footing and interpreting Schrödinger in a way that brought his equation in line with the practical views of the Göttingen physics department. As a proud papa, he expected Einstein to share his enthusiasm and was quite unprepared for Einstein's letter of early December.

"Quantum mechanics is certainly imposing," Einstein granted, "but an inner voice tells me it is not yet the real thing."

Born reports that he took this unexpected criticism "as a hard blow."

"The theory says a lot, but does not really bring us any closer to the secret of the 'Old One.'"

Max Born had known and chatted with Einstein for almost 20 years and yet appears to have spent the rest of his life feeling puzzled by what Einstein meant.

"I, at any rate," Einstein proceeded, introducing what became the most famous metaphor in his canon, "am convinced that He is not playing at dice."

By speaking so vividly, Einstein did his cause lasting damage. Born seems to have pictured God shooting dice and moving pieces across the universal game board and thought that Einstein objected to the randomness of things. Einstein's mistrust actually focused on the alienation between dice throw and action. He found it as unsatisfactory as would a schoolboy who dislikes the board-game version of football. Throwing dice to determine player actions is no substitute for the reality of athletic competition.

In another letter, Einstein told Born, "You believe in God playing dice and I in perfect laws in the world of things existing as real objects." There he made it as clear as language can allow that, for Einstein, the opposite of dice throwing was not a predetermined world, but one in which real things happen for real reasons.

Whenever a revolutionary refuses to accept the revolution's outcome, former comrades-in-arms feel dismayed. The Soviet Union was developing a whole vocabulary to vilify revolutionaries who broke with the party line: right-wing Kautskyite, objectively counterrevolutionary, Trotskyist, and so forth. Nothing that terrible was likely to break out in the world of physics, but Einstein did find himself denigrated and scorned by those who embraced Max Born's probabilistic settlement. Physicists had routinely scorned Einstein's obsessions as windmill tilting and they continued the tradition this time.

"In the course of scientific progress," Heisenberg later wrote, "it can happen that a new range of empirical data can be completely understood only when the enormous effort is made to enlarge [their philosophical] framework and to change the very structure of the thought process. In the case of quantum mechanics Einstein was apparently no longer willing to take this step, or perhaps no longer able to do so."

Although more diplomatic, Bohr, too, dismissed Einstein's objections in a way that relieved himself of any need to wonder whether his critic had a point: "In dealing with the task of bringing order into an entirely new field of experience, we could hardly trust in any accustomed principles, however broad, apart from the demand of avoiding logical inconsistencies." It was a brazen statement for a thinker whose use of complementarity sounded to some like the embrace of inconsistency.

It sounds like Einstein was unwilling to consider the new statistical ideas, but Einstein knew as well as anyone in physics that action often results from chance. His explanation of Brownian motion had shown his very deep understanding of statistical fluctuations based on the perfectly random actions of atoms in fluid, and that great paper had built on previous work in which Einstein developed a statistical approach to unpredictable changes. Indeed, as Einstein fully understood, many classic concepts of physics, like temperature and pressure, have no meaning apart from a statistical one. But he also believed as fervently as Saint Anselm believed in the Almighty that the physical world was real and that physical outcomes result from physical inputs. If Brownian motion can be explained with statistical equations, it is because a real physical action underlies those statistics.

When Einstein wrote of coming closer to "the secret of the 'Old One,'" he stayed with his belief that advances in science link rules to reality. Every step brings the world that much closer to discovering the invisible truth behind appearances. Max Born complained that Einstein had not rejected quantum mechanics "for any definite reason, but rather by referring to an 'inner voice.'" But of course Einstein had immediately given his definite reason: it does not bring us any closer to the secret of the "Old One." He was still determined to see Truth's face.

Born could have responded to that objection: "There is no 'Old One,' no physical givens whose properties and laws are to be discovered. At bottom, instead of nature and natural laws, there is only chaos." But the striking fact is that Max Born never replied to Einstein's objection, never seemed even to grasp what was at stake. Over the years, Born and Einstein had many more exchanges about quantum mechanics, with Born never clutching the nettle.

Born commented on Einstein's initial response to quantum mechanics: "This rejection was based on a basic difference of philosophical attitude, which separated Einstein from the younger generation to which I felt that I belonged, although I was only a few years younger than Einstein." Thus began the myth of Einstein as old fogey. It should immediately provoke suspicion to see how flattering this account is to its teller: all of three years younger than Einstein, Max Born the utterly married Herr Professor of Göttingen presents himself as so much younger at heart than Einstein the philandering man–about–town.

Born responded in a similar tone to another letter in which Einstein said we cannot understand clearly how quantum "machinery" works: "the same argument [about not understanding it clearly] was used by the opponents of the young Einstein. . .". Born slips in the old-fogey-Einstein image as delicately as a prison hit-man slips in a shiv, ". . . who alleged that the consequences of the relativity theory did not make sense."

This linkage of anti–relativity arguments with anti–quantum math became popular, but it misses the seriousness of Einstein's objection. Complaints that relativity made no sense and was incomprehensible usually meant that Einstein thought reality was one way (in which time is relative) while the speaker was confident reality was another way (in which time is absolute). A philosophical opponent of relativity might have said, "I cannot imagine reality working the way you say it works." Einstein's objection to quantum mechanics went a step further, disputing the way the theory had taken reality out of the story altogether and replaced it with a mathematics. Einstein's physics rested on the credo that physical reality exists all the way down to the smallest part and flutter of the universe and that intelligible explanations of what happens must not stray from physical reality's role.

Once, during the war years, Born came close to grasping Einstein's point, "If God has made the world a perfect mechanism, he has at least conceded so much to our imperfect intellect that, in order to predict little parts of it, we need not solve innumerable differential equations but can use dice with fair success. That this is so I have learned, with many of my contemporaries, from Einstein himself. I think this situation has not changed much by the introduction of quantum statistics; it is still we mortals who are playing dice for our little purposes of prognosis—God's actions are as mysterious in classical Brownian motion as in radioactivity and quantum radiation, or in life at large."

How startling it is to find Max Born, even in the 1940s, asserting that thinking about physics was "not changed much by the introduction of quantum statistics." The work of Galileo and Newton had enthroned natural explanation. Something new was afoot when scientists no longer believed that physical events lay behind physical laws.

The one who did grasp the importance of the change was Niels Bohr. His distaste for mathematics meant that he could not follow his colleagues into a world of pure mathematical meaning. He could understand the equations, but that was not where his creative strength lay. This unusual physical imagination might have pushed him out of the quantum story for good, but in his own way Bohr was as restless as Einstein.

Einstein's way was reflected in the letter he wrote Max Born. After saying that God is not rolling dice, Einstein described his own approach to the problem. Its mathematical details are not important because they came to naught, but the approach matters. Einstein was returning to square one, looking for a new way to derive Schrödinger's equation. That was the standard physicist's approach to the mystery of Planck's h—find some new insight into the physical world that leads to Planck's equation. It was also Einstein's approach to gravity where he found a new, relativistic world that led to Newton's predictions. Bohr, however, came at it from the other end. Given quantum mechanics, what can we say about the world?

28

Indeterminacy

When Schrödinger boarded his train out of Copenhagen, neither he nor Bohr seemed to have ceded any ground to the other, but both were shaken. Each had argued passionately for days without winning the other over, and also without finding a reply to every challenge put by the other. Both men could see that they did not fully understand their own ground. Max Born could proclaim himself "entirely satisfied" physicswise, but perhaps that revealed an undemanding nature. Bohr was not satisfied, and after Schrödinger's departure he and Heisenberg worked intently together in an effort to gain a practical understanding of quantum physics through and through.

As Denmark's nights lengthened, the two spent more and more time talking about how the physics worked. They struggled logically through one experiment after another, figuring out its math and what the solution suggested about the physics of the event. Colleagues though they were, and evident as the respect in which each held the other, their imaginations worked very differently. Heisenberg's skill and thinking ran to mathematics. If the equations covered the ground, he would be satisfied.

Bohr wanted to find coherent, unambiguous language that described the logical nature of the case.

Heisenberg thought mathematics provided all the coherence and precision needed.

Mathematics, however, was not the language Bohr wanted to use, or at least formal math was not sufficient. He wanted to know what the numbers were about. The $2 \times 5 = 10$ kind of counting was too empty for him. He preferred the sort that says two sandwiches for each of five schoolchildren yields 10 empty sandwich wrappers. So how are we to understand that p times q does not equal q times p? Max Born was satisfied with a purely mathematical explanation, but Bohr wondered, and he wondered, too, about how we are to talk about quanta if sometimes it is a particle and sometimes a wave? The triumph of the Compton effect had established the correctness of Einstein's idea, but where was its sense?

Heisenberg and Bohr thrashed over these questions, but their styles were too incompatible. It was like trying to compose an opera with music by a formalist like J.S. Bach and libretto by an anti-formalist like Ezra Pound. Each might admire the other's genius, but together they were not going to advance far. At the same time, they stimulated so many thoughts in the other that the moment the two separated, their minds erupted with fruitful ideas.

Bohr left Heisenberg alone at the end of February 1927, when he set off to vacation in Norway, and immediately ideas began to blossom. Heisenberg continued thinking about quantum experiments and how to understand them, but with Bohr gone, Heisenberg suddenly recalled Einstein saying that "Theory determines what we observe," and he decided to see what his theory observed.

Einstein could have told him he was stepping into a logical circle. Heisenberg had said "Here is what we can observe," and he developed a theory to match that. Now he was asking, "Here is the theory, what does it say we can observe?" Not surprisingly, this ride let him off right where he got on board. But the journey was not sterile because it led him to look more deeply into the implications of what he had wrought.

The most important insight was one that has become known variously as the uncertainty principle or the indeterminacy principle. The "uncertainty" term is the less radical sounding of the two. Uncertainty is part of everybody's life, and nobody will be surprised to learn that there are details of an experience that are beyond discovery. "Indeterminacy," however, radically implies there is something in the event outside physical causes. The confusion over terms reflects the deeper confusion over the principle's meaning. Most readers are probably more familiar with the uncertainty term, which was the word Heisenberg used, although many physicists, including Bohr, preferred indeterminacy. This history will take a neutral course and refer to it as Heisenberg's principle.

Such quarreling over interpretation was one of the things science had seemed to abolish. Philosophers had once argued interminably about the interpretation of the most elementary facts, but science after Galileo and Newton made tremendous strides by somehow limiting that tendency. Galileo, like philosophers before him, wrote dialogues in which people disputed the meaning of facts, but once the science of natural law took hold, mathematics and meaning went together. But with quantum physics, from the start, the meaning was elusive. Its laws predicted experience without explaining it.

Planck introduced his hv. What was that?

An element of energy.

Okay, but what exactly is it?

Maybe it is a particle, maybe a wave, maybe a wave packet. Whatever it is, it is omnipresent.

In its mathematical form, Heisenberg's principle states that when you subtract qp from pq you will get, as a minimum, Planck's h. The quantum thus becomes an expression of something—uncertainty, indeterminacy, we cannot quite agree on what it means, but plainly it lies at the heart of the universe.

But what is that heart doing? What does it mean? Philosophically, it puts an end to the hope for a full accounting of quantum events. Heisenberg's principle works like a balance on the scales of justice in which a jury must always find reasonable doubt. If an investigator gets one side of the scale absolutely right, the other side becomes so hazy that the jury cannot know what happened there.

Well, ponders a juror, why not solve one side and get it right and then solve the other side and get it right. Then the investigator will have solved both sides. But that common-sense effort misses the power of Heisenberg's principle. We cannot solve both sides at the same time. Examining the one side permanently obscures events on the other side. Heisenberg said that if we design an experiment to measure both sides of the scale at once, neither answer will be exact. The balance will swing like a doctor's scale gone mad in which the pointer flits up and down and makes it impossible for anybody to determine where reality lies.

Colorful though this image is, it does not get at the physics of the problem. Why must the scale swing so wildly? Heisenberg's paper attempted to give reasons. His essay is generally considered revolutionary. Compared with Newtonian doctrines, it was. From the perspective of the papers in the early days of the quantum revolution, however, Heisenberg was beating a retreat toward normalcy. His most radical early idea—the absence of space and time from quantum mechanics—was now abandoned, replaced by his new principle. In this new paper, things happened at specific times and in definite locations, although we might never be able to discover those exact times and places. Heisenberg also backed away from Max Born's radical probabilities and did not propose to explain his uncertainties in terms of pure chance.

A mathematical view might be that we can know only the probable values of qp, not its exact value, but Heisenberg did look for underlying physical reasons. His most famous example envisioned an experiment in which a physicist tries to look at an electron through a gamma-ray microscope. (Gamma rays are similar to X-rays, but have more energy.) Picking a microscope was Heisenberg's way of thumbing his nose at his theory's vocal enemy, Wilhelm Wien. During the oral exam for Heisenberg's doctorate, Wien had asked Heisenberg about the precision of a microscope's optics? What were the elements that determined how clearly an object could be studied? Heisenberg answered so miserably that Wien wanted to fail him. So Heisenberg was getting a bit of his own back when he asked how you can use a microscope to determine precisely where an electron is. You can peer through the gamma-ray microscope; however, thanks to the Compton effect, the ray that finds the electron and shows its location also knock the electron, scattering it and changing its momentum. Finding out one thing, Heisenberg said, changes and obscures something else.

Heisenberg also used uncertainty to settle the long-standing enigma of the quantum jump. His solution was to say that when energy levels were known very exactly, the time could be known only inexactly. An experiment could say only that the energy change took place over a time of uncertain duration. Thus, the change in electron position no longer looked quite so miraculous.

Finally, Heisenberg's paper beat a retreat from Max Born's radical insistence on probabilities. Heisenberg ended his paper with the remark, "Quantum mechanics establishes the final failure of causality." That sounds plenty radical, but a few sentences earlier Heisenberg had undermined this stance by explaining that the error lies in believing that "When we know the present precisely, we can predict the future." We cannot predict the future because we cannot know the present precisely. That is much tamer than Max Born's doctrine that we cannot predict the future because the present is free to act as it chooses. This backing away from quantum's radicality made the whole revolution much easier for many physicists to accept. But Einstein, who from the beginning said the quantum was "revolutionary," was not to be assuaged by still more cooling. As long as ψ, the key element of the new quantum mechanics, had no real-world meaning, he was not going to quiet down. Einstein wanted to press on with the revolution and find new concepts that would explain what psi really was.

When he finished his paper, Heisenberg dashed a copy off to his comrade in calculations, Wolfgang Pauli. Not surprisingly, Pauli thought it was great. After all, it was Pauli who has pointed out that uncertainty is implicit in the math of quantum mechanics. Then Bohr returned from the chills of his Norwegian winter, excited by the thoughts he had developed beneath the northern lights. He was startled to learn that his assistant had been busy, too, and had already submitted an article to the leading journal of quantum physics, *Zeitschrift für Physik*.

To Heisenberg's dismay, Bohr was displeased with the new paper, finding it wrong in some of its physics, often superficial, and insufficiently radical. Bohr in Norway had concluded that quantum physics stands in perpetual contradiction to classical physics, and he was most unhappy to see a paper from his institute hide this decisive contradiction behind a screen of uncertainty. A new, ferocious argument erupted between Bohr and Heisenberg. Its centerpiece was Bohr's insistence that Heisenberg withdraw his article so that it could be revised to agree with Bohr's own pending article. Heisenberg refused to change a word.

Perhaps it was during this ordeal that Heisenberg developed his sympathy with the harangue Schrödinger had endured. The ceaseless demand for surrender came not merely from a senior colleague or admired genius, but from Heisenberg's hero and employer. Could he resist such pressure forever?

What was wrong with the paper?

For one thing, the experiment with the gamma-ray microscope was wrong. As he had during his oral examination, Heisenberg again missed the subtleties of the system. The limitations in Heisenberg's equation were correct, but not for the mechanical reasons Heisenberg had given.

What were the reasons then?

Bohr could not explain himself. His instincts told him that Heisenberg's physics was too clumsy and that the real reason was deeper, less mechanical. The paper, Bohr insisted, was often superficial.

Heisenberg thought not.

Striking examples of superficiality lay in the article's many references to Einstein. The name pops up throughout the essay, starting with his second sentence, and Heisenberg specifically sought to compare his work with Einstein's, boasting that "it is natural [in its uncertainty relation] to compare quantum theory with special relativity." Heisenberg then went on to compare Einstein's idea that nothing can be absolutely simultaneous with his own notion that position is uncertain.

Heisenberg liked the claim, but it missed Einstein's achievement of removing ambiguity from his account of physical phenomena. Bohr had an idea for removing all ambiguity from quantum phenomena as well, and he wanted Heisenberg to hold off publication until that idea could be included.

Heisenberg was revising nothing. Was there nothing in the paper for Bohr to admire?

The moments of untamed radicality remained admirable in Bohr's eyes.

The radicality of the uncertainty mathematics itself was good. It asserted that you could never see the whole of a phenomenon at once. If p and q referred to Newtonian concepts of position and momentum, you could not get a complete Newtonian picture of quantum behavior. Or if p and q symbolized thermodynamic notions of time and energy, you could not get a complete thermodynamic picture of quantum change. This limited view of reality fit well with Bohr's epiphany in the Norwegian wood.

Bohr also liked the kind of thinking that led Heisenberg to write "the 'orbit' [of an electron] comes into being only when we observe it ... [it] can be calculated only statistically from the initial conditions." This kind of antirealism appealed to Bohr very much, and doubtless he also found plenty to enjoy in Heisenberg's remark that although a person might argue "that behind the perceived statistical world there still hides a 'real' world in which causality holds, ... such speculation seem to us, to say it explicitly, fruitless and senseless."

Einstein would have replied that the real world can be found through new concepts that show the connections between statistical observations, but Bohr did not think that way and wanted no changes in that approach.

Mostly, however, Bohr did want changes in the paper and Heisenberg firmly refused to make them.

This dispute could end only in ugliness or with intervention from the gods. In this case, the gods lowered a rescuer into the scene. Pauli suddenly arrived in Copenhagen for a visit, and both men fell on him with their accounts of the disagreement.

After listening to both sides Pauli told them they were not in dispute.

Then what had they been fighting about?

Heisenberg, Pauli said, had presented an account of the physical meaning of quantum theory that agreed with Bohr, but Bohr had pushed the idea to greater depth. If only each of them could calm down and understand the other, they would both see that there was no fundamental disagreement between them.

It was the great tactic of conflict resolution—find a point of common ground and declare it to be the central ground.

Heisenberg did not pull his paper from publication, but he did add a passage to its closing. He retreated from a few positions, notably, admitting that the gamma-ray experiment is "not so simple as was assumed."

Mostly the effect of Heisenberg's addendum was like the "to be continued" ending in a serialized Charles Dickens episode. Much wonder and amazement was inspired by the chapter in print, and yet it served largely to make readers pace eagerly in anticipation of what would come next. "I owe great thanks to Professor Bohr," Heisenberg concluded, "for sharing with me at an early stage the results of these more recent investigations of his—to appear soon in a paper on the conceptual structure of quantum theory—and for discussing them with me."

With these words, Heisenberg put the quantum revolution on hold until Bohr's paper should "appear soon." And everybody knew that Bohr was a slow writer.

29

A Very Pleasant Talk

Another Solvay Conference was scheduled for late in October 1927. Einstein had told Lorentz he never wanted to be invited again, and the French and Belgians still despised Germans, but the conference could never live up to its ambitions without a German presence. Modern physics *was* German physics. Planck's declaration immediately after his country's defeat—"no foreign or domestic enemy has taken ... the position German science occupies"—had proven correct. The world wanted to spurn them but could not. Already in the spring of 1926, Lorentz had brought the matter to the attention of Albert, King of the Belgians. Albert had amazed Europe by standing up to the Germans and acting more like a monarch from the days of Henry V than today's constitutional symbols. He had ample reason to begrudge Germans their recovery. Ever the diplomat, Lorentz spoke his piece tactfully enough to report that he and the king saw eye to eye: "His majesty expressed the opinion that seven years after the war, the feelings which they aroused should be gradually damped down, that a better understanding between peoples was absolutely necessary for the future, and that science could help to bring this about. He also felt it necessary to stress that in view of all that the Germans had done for physics, it would be very difficult to pass them over."

Germany was clawing its way back onto the world stage. By 1927, the economic hardships of the first half of the 1920s were merely a bad memory, and Berlin had become a center of artistic achievement. Prominent visitors paid call on the city. The American composer George Antheil had seen Berlin in its ruin but returned during this "golden" period and reported, "The electric lights were back in their sockets. The red carpets, new ones, were down on the floors of the expensive hotels. People had their brass doorknobs out again." He did not mention that among the posters for films and cabarets was a blood-red one that proclaimed, "The bourgeois state is coming to an end. A new Germany must be forged." Goebbels was getting down to work.

Freud, the only thinker able to rival Einstein's fame, came to Berlin and met Einstein. "He understands as much about psychology as I do about physics," Freud reported, "so we had a very pleasant talk." The weather and music were two neutral topics in which both men were expert. And of course each thought he understood more of the other fellow's field than he did.

Einstein's favorite psychological doctrine came from Schopenhauer: We can do whatever we want, but we cannot choose our wants. Freud's ideas owed much to Schopenhauer and were quite in keeping with that sentiment, although as a medical man, he attracted patients by offering the hope that they could overcome their neurotic actions. Meanwhile Freud thought of himself as every bit the scientist that Einstein was and he based his imagery of how the mind worked on the mechanist's faith that there is a natural reason for everything that happens. Einstein might have found Freud's steam-engine mechanics naïve. Possibly he laughed at Freud's notorious insistence that a patient's refusal to accept his diagnosis was merely one of the patient's symptoms, but Einstein agreed entirely with Freud that every event has its reason.

That was the notion he took with him to Brussels and the 1927 Solvay Conference, the greatest gathering of physicists since the first Solvay in 1911. The official conference photograph shows the century's grandest collection of physics heroes. Einstein radiates the presence of a movie star. Arthur Compton sits right behind him. Lorentz has the chair beside Einstein. Marie Curie is there. Bohr is there. Rounding out the German delegation are Heisenberg, Planck, and Max Born.

Solvay Conference, 1927. *Front Row:* Irving Langmuir, Max Planck, Marie Curie, Hendrik Lorentz, Albert Einstein, Paul Langevin, Charles Guye, Charles Wilson, Owen Richardson. *Middle Row:* Peter Debye, Martin Knudsen, William Bragg, Hendrik Kramers, Paul Dirac, Arthur Holly Compton, Louis de Broglie, Max Born, Niels Bohr. *Back Row:* Auguste Piccard, Emile Henriot, Paul Ehrenfest, Edouard Herzen, Theophile de Donder, Erwin Schrödinger, Emile Verschaffelt, Wolfgang Pauli, Werner Heisenberg, Ralph Fowler, Louis Brillouin.
Credit: Photographie Benjamin Couprie, Institut International de Physique Solvay, courtesy AIP Emilio Segré Visual Archives.

The conference's official theme was "Electrons and Photons." At the previous conference, in 1924, the existence of light quanta had been, at best, controversial. Now, rebaptized as photons, they were accepted by everybody in the room. A new controversy bedeviled physics. Were quantum events caused? Lorentz, as always, chaired the sessions, and in his opening remarks he asked, "Could not one maintain determinism by making it an article of faith? Must one necessarily elevate indeterminism to a principle?"

Scientists have their doctrines, but they rarely assemble them into some sort of Nicene Creed whose tenets are duly recited and accepted. Lorentz's proposal went nowhere, and yet his attitude was held by many of the Solvay participants. The two heroes of X-ray studies, William Bragg and Arthur Compton, gave papers and both of them leaned toward the explanatory school; that is, they discussed experiments and looked for physical causes to explain why what happened happened. De Broglie and Schrödinger also sat in this camp. Unfortunately, de Broglie tried to go beyond just believing in explanations and actually proposed one. A few months earlier he had suggested some subtle ideas. His original notion of wave-particle harmony singing a duet had also been evocative, but in Brussels he suddenly reverted to the abandoned concept of a "pilot-wave" guiding the photons. Nobody championed his proposal, and in the confusion everyone, apparently including de Broglie, forgot that there was another de Broglie explanation out there. It was only generations later, well after Einstein's and Bohr's death, that some philosophers of science looked at de Broglie's work and found it provocative.

The opposed camp did not try to explain phenomena so much as describe them. Born and Heisenberg gave a joint paper in which they insisted that although quantum events could be described, finding reasons for the events was impossible. They ended their lecture with the tossing of a gauntlet, "We maintain that quantum mechanics is a complete theory; its basic physical and mathematical hypotheses are not further susceptible of modifications." If the explainers were hesitant to adopt Lorentz's credo, the describers were eager to join the prophets who claimed their teachings to be the final revelation on the matter. This was Max Born's reply to Einstein's assertion that quantum mechanics was "not yet the real thing." He said it was not only the real thing; it was the final thing.

Until that moment Einstein might have viewed the continuing struggle over quantum physics as part of science's normal work. You sweat. You learn. You get transitory ideas. You follow false leads. You finally take a stride or two that ties the subject to the rest of physics and makes the whole thing comprehensible. In that context, quantum mechanics could be an impressive achievement even if it was not yet the real thing. But Born and Heisenberg were now saying that the work was done, the revolution over. It was never going to be more intelligible or more fully connected to the rest of physics.

Starting with that Solvay Conference of 1927, Einstein ceased to be merely skeptical or dissatisfied with the state of quantum physics. He became defiant, refusing to bow before the claim of completeness, finality, and unmodifiability. Quantum completeness became for him the new ether, a doctrine to be resisted, precisely because it put the undiscoverability of proof as the ultimate proof. And the notion that something must be forever undiscoverable was offensive to him. The reason that all objects fall at the same speed no matter what they weigh had, for centuries, been unknown and beyond all discovery, yet one afternoon the reason had, in a flash, been understood.

At the close of the formal sessions, Einstein had his first opportunity to show his new mood. Bohr gave his summary of what physics was to be, now that mathematical probability had replaced physical causes. This was the long-awaited new thinking promised at the end of Heisenberg's paper. Bohr had presented these same ideas a month earlier at a conference in Italy, but many physicists, including Einstein, had not attended that affair and did not know what Bohr had said.

Bohr was never an articulate man, and any attempt to summarize his remarks inevitably makes them seem clearer and more focused that they appeared to his audience. As he spoke that autumn afternoon in Brussels, the physicists peering through the smoky air could collect only parts of what he was saying. Einstein tended to see wider and deeper than his colleagues and appears to have grasped the nub of Bohr's point. However, for many it was undoubtedly like trying to follow the voice in the babbling brook. One thing was plain; the kind of solution Einstein had anticipated in his Nobel lecture was not Bohr's. Einstein had imagined a solution in which relativity was a limiting case; Bohr presented something quite different.

The very nature of the quantum theory, Bohr told his audience, forces us to take the distinct concepts of classical physics and use them as complementary, exclusive features in a description of quantum phenomena. Physicists have been trying to understand light's wave and particle nature as a whole. It cannot be done. In some experiments, light appears to be a particle; in others, it appears to be a wave. The concepts are complementary, meaning we can understand light's nature only if we abandon the search for a coherent unit and think of it as sometimes being like a particle and sometimes like a wave. We can think this way without becoming contradictory because these complementary concepts are also exclusive. That is, we can never perform an experiment in which light's wave and particle natures are both measurable. This complementarity is the profound physical meaning of Heisenberg's principle. Whatever p symbolizes in Heisenberg's algebra is the complement of whatever q symbolizes. Thus, *position* and *momentum*, or *time* and *energy* are complementary and exclusive notions in quantum physics, even though they are coherent concepts in traditional physics. Therefore, it is impossible to design an experiment that will simultaneously determine a photon's position and momentum, or its time and energy.

Bohr was not given to mathematical reasoning of the sort that Max Born had used in arguing that a quantum wave was a mathematical fiction describing probabilities without underlying causes. Nevertheless, Niels Bohr's theory of complementarity stood firmly in Max Born's camp, because it insisted that experimental outcomes could only be described, not explained in terms of real causes. Bohr was replacing Einstein's theoretical arch in which facts and concepts supported one another with a new style in which rival facts and rival concepts tolerate each other. Any observation of atomic phenomena involves an interaction between the phenomenon and whatever tools the scientist uses to observe the event. Thus, Bohr told his audience, "an independent reality in the ordinary physical sense can neither be ascribed to the phenomenon nor to the agencies of observation."

The logic of Bohr's argument rested on one novel dogma: the language of classical physics could not be abandoned. Quantum mechanics had to be understood in classical terms and, in order to avoid seeing classical terms like wave and particle as contradictory, they had to be viewed as complementary. Only when taken together did quantum mechanics "offer a natural generalization of the classical mode of description." Bohr gave as a reason for this axiom of complementarity that "every word in the language refers to our ordinary perception." Classical physics does an excellent job of describing the events we can study with our eyes, ears, and other sense organs. Our language, thus, describes experiences that are well suited to a classical understanding. Quantum phenomena, however, cannot be tracked directly with our senses. We see hints of events—changes in atomic spectra, for example—but when we try to talk about them we find that we must think in classical terms, and in the quantum world such thinking makes sense only if we understand them as complementary notions rather than contradictory ones.

But surely, some readers are bound to object, we can make up a new language that is more suited to this new world.

Physics had already done that by using the language of mathematics. It had coined a whole string of symbols that applied to quantum events: h, $h\nu$, p, q, and ψ. They work perfectly well in equations, the sentences of mathematical language. It is just that when we try to understand these symbols—that is, when we try to imagine what is happening—we fail. Our perceptions have trapped our imaginations in a classical world where quantum events do not apply.

This argument aimed at a secret assumption behind Einstein's talk of "accessibility." He often said that the world was lawful and the laws were accessible to us. By that last point he meant that logic and fact were sufficient for us to discover and understand natural laws. But Bohr was saying that accessibility rests on a third thing. In addition to facts and logic there was language, and language can never dive down from the classical to the quantum level. Rather cheekily, Bohr claimed an alliance with Einstein in his attitude toward language, "We find ourselves here on the very path taken by Einstein of adapting our modes of perception borrowed from the sensations to the gradually deepening knowledge of the laws of Nature."

Einstein did not see Bohr as a true and faithful disciple, and he never had any patience with Bohr's complementarity. It seemed to him that it was simply giving up on the effort to understand the physics of quantum events. A similar "solution" could have followed the Lorentz equations if Einstein had tumbled over the edge of the Alps and never completed his work. Lorentz's equation was demonstrably correct. How then could anyone explain the electricity-magnetism paradox that Einstein so worried over? Surely, no one would have settled for the "explanation" that the very nature of electrodynamics forced scientists into understanding electricity and magnetism as complementary concepts.

As soon as the general discussion began, Einstein rose to show how well he understood the arguments that he rejected. He pointed out a contradiction between Max Born's probabilistic understanding of quantum mechanics and Niels Bohr's complementary interpretation. Einstein's arguments were always grounded in experiments, usually imaginary ones. In this case he pictured a photon passing through a slit to a photographic plate. This thought experiment happens in reality every time a photographer snaps a shutter, or, going back further in history, is inherent in pinhole image studies that had determined that light is out in the world and not part of the eye. Light bends when it passes through the slit. Schrödinger's equation shows that this bent light might land anywhere on the film. Max Born's probabilistic interpretation says the chances of the photon's striking a point are spread over practically every point on the film. When the photon strikes the film, however, the probability of its landing anywhere else drops to zero. This result was in keeping with Max Born's indeterministic viewpoint. We cannot know what will happen until it has happened.

Bohr took a stronger view and insisted that during the undetermined period the photon maintained a virtual reality as though it were smeared like butter across every point where it might strike. This idea, however, contradicts relativity because the collapse of the smeared photon into a single point would be instantaneous (faster than the speed of light) and, like Newton's discredited account of how gravity worked, would constitute action at a distance.

Einstein proposed an alternate explanation. The photon passes through the slit and follows a distinct trajectory to the point on the film where it strikes. Quantum mechanics cannot describe this trajectory and is, therefore, not the whole story. It was another William Tell bulls-eye. Einstein had gotten, at first hearing, to the central mystery of the quantum revolution: the quantum collapse. What happens when a quantum particle that might appear in many possible places appears in one particular place? Einstein did not know, but he had heard enough to see that the revolutionaries did not know either.

Bohr appears to have had no immediate answer. Heisenberg, Pauli, and Dirac, however, were more concerned with saving quantum mechanics than with complementarity, and they insisted that Schrödinger's wave equation does not represent actual events in space-time. It expresses what we observers can know of events. We know the photon entered the slit and that it hit the film at some point. The wave equation tells us what the chances were of its hitting any point. True, quantum mechanics cannot describe what happens between any two observations, but the incompleteness cannot be held against the theory. It is built into our own experimental limitations. We can only perform experiments that give us part of the data. We can never get the full, objective data; therefore, we cannot develop a full, objective theory. Quantum mechanics is as complete as our experimental limitations will ever let a theory be.

This answer was not to be the last word. The dispute over the meaning of the quantum collapse, or whether there is even any such thing, continues to this day. It was typical of Einstein to have seen at once the loosest thread on the sweater and to have given it a good tug. It was typical, too, of people caught in the midst of revolution to have missed the importance of what had happened before their eyes. Einstein had transformed Bohr's complementarity from being an unambiguous account of what happens in quantum events into being one way of talking about events. Only over time would physicists feel the hurt from this sting. Complementarity would become know as an "interpretation"—the "Copenhagen interpretation"—rather than as a theory, and the worldwide shorthand for the paradoxes of quantum theory would not be Bohr's complementarity but Heisenberg's principle.

Not that Einstein wanted to prop up Heisenberg. His thought experiment with photon and film had not challenged Heisenberg's principle, but now Einstein did turn his attention there. He began looking for an experiment that would allow a more complete collection of data than the Heisenberg team thought possible. If he could find a technique that allowed the simultaneous discovery of position and momentum or time and energy, he would prove that quantum mechanics had indeed not yet brought us to the secret of the Old One.

This effort led to the most famous set-pieces of the Einstein "debate" with Bohr over quantum mechanics. Einstein, Bohr, and Ehrenfest would meet in the hotel dining room for breakfast. Einstein would propose a thought experiment. Bohr would think about it. Ehrenfest played the role of silent spectator, like an onlooker at a chess game between two grandmasters, but along with a sports fan's fascination, he felt the horror a child knows when parents enter into a quarrelsome divorce. One side might seem to be right, but even so, how can someone who loves them both choose?

Many of Einstein's thought experiments concerned photons passing through slits. A typical one used a shutter to guard the slit. The shutter opens. Light passes through and strikes a film plate. As it passes through the slit, the light strikes the moving shutter and then strikes the plate. If we study both the plate's and the shutter's reaction to the photon collision, we can calculate the time, position, and momentum of the photon to a greater accuracy than the uncertainty principle allows.

Einstein's mistake in this reasoning was to overlook the inherent uncertainty in the experiment. Suppose the mass of the plate is infinite. Then the photon's momentum will not knock the film plate ever so slightly. Instead, according to the physics of momentum, the plate's mass will increase slightly, except that changes in infinite things cannot be measured. The experiment would give no data. So we will assume the more realistic condition that the film plate's mass is not infinite. Now the photon's momentum will jar the film plate ever so slightly, allowing (in principle) a technician to measure the photon's momentum; however, jarring the photographic plate will slightly blur the point where the photon landed. Working through the math in detail shows that the Heisenberg principle survives intact. The more certain the momentum, the less certain the point of landing. Similar considerations show that measuring the shutter action also includes enough uncertainty to make it impossible to calculate the exact time and energy of the photon's interaction with the shutter.

A certain kind of philosopher would have objected that this reply "begs the question." It assumes the point that it set out to test—that uncertainties in the photon's interaction prove the validity of Heisenberg's principle. Einstein raised no such objection, however, because scientific arguments are settled by experiment, not by logic. Einstein used logic simply to look for contradictions. In this case, Heisenberg's principle was consistent even if Einstein did not think it was coherent. So he moved on, looking for other experiments.

During the day's program at Solvay, Heisenberg and Pauli would analyze the experiment that Einstein had proposed. They would find some point where the uncertainty principle fought back, and over dinner, Bohr would refute the experimental effort while Ehrenfest looked on.

At the end of this ordeal both Heisenberg and Pauli felt elated. They left Solvay convinced, as Heisenberg reported in a letter to his parents, that they had won. "I am satisfied in every respect with the scientific results. Bohr's and my views have been generally accepted; at least serious objections are no longer being made, not even by Einstein and Schrödinger."

Heisenberg was getting carried away. Others knew that doubts persisted, especially among Einstein and Schrödinger. Ehrenfest sent a more reliable letter back to his students giving a chatty account. "BOHR was towering over everybody. At first not understood at all . . . then step by step defeating everybody. Naturally, once again the awful Bohr incantation terminology. Impossible for anybody else to summarize."

Ehrenfest's tone, simultaneously awestruck and irreverent, suggests why he was so popular and effective a teacher. His letter also included a striking bit of news, "Every night at 1 a.m. Bohr came into my room just to say ONE SINGLE WORD, until 3 a.m." You can bet Einstein was not out lobbying all night.

Ehrenfest also reported on the Einstein-Bohr debates. "It was delightful for me to be present during the conversation between Bohr and Einstein. Like a game of chess, Einstein all the time with new examples. In a certain sense, a sort of Perpetuum mobile of the second kind to break the UNCERTAINTY RELATION. Bohr from out of philosophical smoke clouds constantly searching for the tools to crush one example after another." Evidently Ehrenfest did not realize that Bohr was like a chess grandmaster with powerful coaches who had studied the position and proposed counter moves.

"Einstein, like a jack-in-the-box," Ehrenfest continued, "jumping out fresh every morning. Oh that was priceless. But I am almost without reservation pro Bohr and contra Einstein. His attitude to Bohr is now exactly like the attitude of the defenders of absolute simultaneity towards him." Perhaps he should have pointed out, however, that the defenders of absolute simultaneity did not understand Einstein and did not argue by producing a series of thought experiments designed to test Einstein's concept.

Einstein, for his part, was exhausted. After the conference, with Louis de Broglie aboard the train from Brussels to Paris, Einstein said he was getting old and that a younger man might do better. "Carry on," Einstein told de Broglie, "You are on the right road."

But youthful de Broglie immediately abandoned the effort while Einstein slogged on.

30

The Dream of His Life

When physics became this baffling, Einstein frequently turned to a new woman, and he had a candidate ready for the role. She was Toni Mendel, a well-to-do widow, Jewish, about his age. She would arrive at his home in her chauffeured limousine to whisk him away from Elsa's company to a night at the opera. But this time, frustrating as Solvay had been, the story took an unexpected turn. Einstein collapsed on a visit to Switzerland. He was carrying a suitcase up a snowy hill when he fell straight down and lay there beside his host, a German industrialist named Willy Meinhardt. He had insisted on hauling his own bag. Meinhardt was terrified that he had allowed his famous guest to bring on a heart attack. Slowly he got Einstein off the ground and even more slowly the pair made their way to the Meinhardt chalet. There had been no heart attack, but Einstein was plainly exhausted. Perhaps, too, he was worn down. Solvay had been hard and then just a month before his own collapse, on February 4, 1928, Hendrik Lorentz had died.

"I have behind me a good and wonderful life," Lorentz had told his daughter Geertruide, shortly before he began showing the effects of his final illness. Einstein went to the funeral as the representative of the Prussian Academy. The ceremony itself was an enormous affair with astonishing touches of national honor. The Dutch telegraph system observed three minutes of silence, starting at noon. Government

276

buildings in Haarlem—the Town Hall, railway station, and cavalry barracks—flew their flags at half-staff. A special train from Leiden brought students and teachers to the funeral. A photograph of the funeral cortege passing through Haarlem's main market square shows that it was packed with ordinary citizens, the men baring their heads, to pay their respects to the great scientist. At the graveside Ehrenfest, Rutherford, Langevin, and Einstein each spoke about Lorentz. "The noblest man of our times," Einstein called him. He spoke again the next day at a memorial service held in Leiden.

How deeply the spectacle moved Einstein remains unclear. He once remarked to an assistant as they walked in a funeral procession, "Attending funerals is something one does for the people around us. In itself it is meaningless. It seems to me not unlike the zeal with which we polish our shoes every day just so that no one will say we are wearing dirty shoes."

A few weeks later he was sprawled in the snow. When he was fit enough, Heinrich Zangger, his old friend from the University of Zürich, came down and oversaw his return to Berlin. Elsa took charge when he arrived. Ten years earlier, she had used her nursing skills to good effect, both for Einstein and herself. At the end of that period Einstein's health was recovered and he had agreed to divorce Mileva. Now he was back in Berlin, badly in need of steady care.

Elsa was ready. She once told a friend, "God has put so much into him [her husband] that is beautiful, and I find him wonderful, even though life at his side is debilitating and difficult in every respect." Besides looking after him, Elsa also seized the moment to bring an ally into the house. This was Helen Dukas, a woman whom Elsa personally selected to be Einstein's new secretary. She was to become a lifelong assistant and, after his death, coexecutor of Einstein's literary estate. She seems to have been beyond seduction, for there is no evidence that Einstein ever made a move in her direction. With her arrival, Elsa at last had a permanent wall against a long-open flank.

Einstein's society doctor, Janos Plesch, combined flair with knowledge. John Maynard Keynes pronounced him "something between a genius and a quack." Plesch ordered Einstein to relax, stay at home, stop sailing his boat, and minimize contacts with people beyond the family. Elsa stood watch by the front door, turning away most of the would-be visitors. When visitors did come in, she soon chased them out to make sure they did not overtire the invalid. She led him away for a long summer on the Baltic beach where he wrote letters, read Spinoza, and thought about physics. That summer turned out to be the last grand one that Berlin would enjoy for many decades, and Einstein missed most of it.

In May, a national election led to a new coalition government, one that tilted left. In those days, leftist German leaders were the ones who fully accepted the legitimacy of the republic and were neither nostalgic for monarchy nor ambitious for tyranny. Nationalist parties appeared to be in retreat from popularity. Brecht's proletarian production of *The Threepenny Opera* was the "show of the season," suggesting that the city's bourgeoisie boasted a left sensibility. Seventy plus years later, it is impossible to view Germany in the late 1920s and not see the shadow of coming catastrophe, but the future's shadow is seldom visible to those on the scene. Count Kessler's diary for that summer shows the opposite of worry. *14 June 1928: Lunched with [the composer] Richard Strauss . . . [who] aired his quaint views, about the need for dictatorship, and so on, which nobody takes seriously.* Republicans were so confident that summer they could laugh.

Meanwhile, Einstein was sunbathing on the Baltic shores, enjoying a relaxed correspondence with Ehrenfest and Schrödinger. A photograph taken that summer shows him lounging on a beach chair and wearing a robe that was sufficiently eye-catching to be a stage magician's costume. Women's bathing suits had changed radically, making the seaside a perfect place for Einstein to enjoy himself while scribbling letters and postcards. He sounded like someone enjoying his convalescence. "Here I am forced to laze about under splendid beech trees," he told Ehrenfest. "How happy one can be in quietude and solitude. It is also wonderful for contemplation. . . . I now believe less than ever in the essentially statistical nature of events and have decided to use what little energy is left me in accordance with my own predilection, regardless of the present hustle around me."

Did Ehrenfest shriek when he read that? When had Einstein ever surrendered his predilections to the hustle around him? Whatever was heroic or petty about him came from his ability to ignore all hustle and follow his own compass. He must have been truly exhausted if he ever thought he might just go along with the ideas about quantum statistics. Ehrenfest needed help himself just then, precisely because he had lost faith in his compass. He had sided with Bohr and rejected Einstein's opposition to quantum mechanics, but he was not happy. He feared that physics no longer made sense to him, or at least that he could not use it creatively. This notion seems to have been part of a depressive delusion. The physicist and "father of the H-bomb," Edward Teller, was a student in those days and recalls, "Ehrenfest did not invent quantum mechanics, but he had the reputation [among students] of understanding it, or at least of explaining it, better than those who invented it."

The real inventors of quantum mechanics that summer were distracted by more routine work. Heisenberg had just arrived in Leipzig where his reputation and up-to-date physics had been summoned to restore the university's program. Bohr was teaching too, although one must never forget that teaching for Bohr was never routine. He had become one of the most persuasive teachers since Plato; it was how he made his mark. In August 1928, as an example, an unknown student named George Gamow turned up at the door of Bohr's institute and asked for an audience, which, grudgingly, was granted.

What are you up to, Bohr wanted to know.

Gamow reported that he had been studying a kind of radioactivity known as alpha-decay. Rutherford had won his Nobel prize for studying the phenomenon.

What about alpha-radioactivity? Bohr wondered.

Gamow had shown that such radiation was not the result of some mysterious atomic instability, but a logical consequence of quantum mechanics.

Ah. "My secretary tells me you cannot stay more than one day because you have no money," Bohr said.

Gamow was more like a poet than a physicist and could not even afford to buy food. He had bummed his way to Copenhagen that summer in the same bohemian spirit that had sent William Faulkner exploring Paris.

"If I arrange for you a fellowship," Bohr wondered, "would you stay for a year?"

"Yes," Gamow told him and Bohr won another disciple.

Einstein took advantage of the long, enforced calm to think of physics, mathematics, and the laws of nature. He wanted an experiment that would show the incompleteness of quantum mechanics, but there he was like a chess champion looking for a move that had been missed in the heat of surprise and competition. He had plenty of time before the next match and his letters that summer show that he was not looking for an immediately publishable way to recoup against quantum mechanics.

While Einstein recovered, Schrödinger called for a continuation of the revolutionary effort. He wrote Bohr a letter saying that the classical concepts represented in the Heisenberg principle by p and q would eventually have to be replaced by less vague ideas.

No, Bohr said, the old concepts could survive perfectly well in the framework of complementarity. Note, by the way, that in this exchange it is Bohr who is insisting on accustomed principles while Schrödinger is calling for a more radical break with the old ways of thinking. Yet this paradox does not mean Schrödinger was the real revolutionary. Schrödinger and Einstein had both expected a traditional revolution— if we can use that oxymoron. Einstein's achievement showed what a traditional revolution could accomplish. It had redefined space, time, and gravity (the revolutionary part) but kept it all in a context of natural explanations (the traditional part). Newton would have recognized and sympathized with Einstein's ambition. Bohr and Max Born, however, had produced a revolution in which the terms kept their old meanings, but they were removed from physically coherent context into one of pure mathematical description. In that sense, it is fair to compare Einstein and Schrödinger with freethinkers during an old regime who had wanted great changes but were horrified when the king himself was sent to the guillotine.

From his seaside beach chair Einstein kept up with the Schrödinger-Bohr letters and he sympathized. "Your claim," he wrote Schrödinger, "that the concepts p, q will have to be given up if they can only claim such 'shaky' meaning seems to me to be fully justified." (It should be reassuring to nonphysicists who scratch their heads over just what p and q mean to see that trained, creative physicists also muttered in dismay.)

Einstein continued, telling Schrödinger, "The Heisenberg-Bohr tranquilizing philosophy—or religion?—is so delicately contrived that for the time being, it provides a gentle pillow for the true believer from which he cannot very easily be aroused. So let him lie there."

However, at some point Einstein did find a challenge that he could set before Bohr at their next meeting. When he hit upon his secret retort is uncertain. Whenever he thought of it—the night after Solvay 1927, the night before Solvay 1930, or somewhere in between—he would be ready with his zinger. During the wait it was the search for a single field uniting gravitational and electromagnetic fields that held his thoughts.

He had been struggling with this question throughout the 1920s, but it now seemed a bit of an old-fashioned problem. Between Einstein's post-Nobel lecture at Göteborg in the summer of 1923 and his convalescence on the Baltic five years later had come the whole of the quantum revolution: matrix mechanics, wave mechanics, Max Born's probabilistic interpretation, Heisenberg's principle, and Bohr's notion of complementarity. The effect of this intervention was to divorce parts of physics in a way that had previously seemed impossible. The reluctant divorcee was Einstein's physics—general relativity, a set of mathematical laws whose physics was governed by space-time, and from which the present flows continuously from the past. Georges Lemaître had only just then begun developing an idea based on Einstein's laws that would become known as the "Big Bang" theory of the universe; it is the perfect model of Einstein's physics. Once, billions of years ago, something happened and since then the universe has been evolving like a wondrous symphony in which each astonishing note flows from its predecessor. This cosmological physics is objective, coherent, and inevitable.

Quantum physics—mathematical, discontinuous, and probabilistic—was the breakaway partner. If the heavens since the Big Bang are the model for macro-physics, the Big Bang itself is the model for quantum physics. On the two sides of it you have two distinct, unrelated states of nature; between them lies the mysterious discontinuity, the incomprehensible verb whose essence is invisible but whose intervention changes everything. By persisting with his search for a unification of gravity and electromagnetism in a continuous, inevitable law of nature—a law that did not even include the quantum *h*—Einstein was defying the very revolution he had once prophesied.

As he convalesced, Einstein suddenly had one of his most famous ideas. As usual, he was testing one of his physics' most fundamental assumptions. In this case he was reexamining the Riemann geometry that supported his general theory of relativity. Particularly he was challenging its most distinctive feature, its absence of parallel lines in anything but infinitely small spaces. In this search he really had gone back to the starting point, because Riemann's no-parallelism is what makes everything in the universe appear to curve as it moves through space. No-parallelism was the aspect of Einstein's theory that guaranteed that the relative time for every moving object is slightly different. If any of the anti-Einsteinians had been up to it, the way to defeat relativity and reduce Einstein to the level of one more overphotographed celebrity was to overthrow its geometric basis. Of course, none of them were up to it, but there was Einstein, picking at it himself.

In June 1928, Max Planck appeared before the Prussian Academy and presented Einstein's new mathematics of "distant parallelism." (Einstein was not yet strong enough to read his own paper.) In this geometry, parallel lines were possible, although they would meet in infinitely large spaces. Somehow this doctrine became part of the Einstein legend: He was the smartest man in the world; only 12 people could understand his theory; time is the fourth dimension; and parallel lines meet in infinity. Those details would become what the well-educated schoolchild knows of Einstein.

Distant parallelism was a purely mathematical idea, but Einstein suspected that it could provide the basis for a new physics step that would at last allow him to unite the gravitational and electromagnetic fields. As summer turned to autumn, his confidence of success grew. Elsa told a friend, "He has solved the problem whose solution was the dream of his life."

It was while Einstein was in that good mood that the *New York Times*'s Berlin correspondent paid a surprise visit to the Einstein home. He gave no record about why he dropped by, but probably he was just following the routine of a good beat reporter who periodically checks out his usual suspects. So the newspaperman, Paul Miller, knocked on Einstein's door and Elsa opened it. Her job as gatekeeper demanded that she shoo off unwanted visitors, but Miller had not become a leading reporter by being shy. He persuaded Elsa to permit an audience with the resident genius.

"He will scream and shriek and rave," she advised, leading Miller one flight up to Einstein's study, "but never mind that. When he calms down he may have something to say."

Elsa opened the study door and found Einstein so absorbed in doing mathematical manipulations on a writing pad that he did not look up. Books were scattered about. Einstein remained lost in his work on unified field theory until Elsa called to him.

As promised, Einstein began to curse the interruption.

Miller immediately called out a question about the proposed new telescope that was eventually built for Mount Palomar, "Professor, do you think the new telescope with a 200-inch lens to be erected in California will be able to disclose visual proof of relativity? You are aware that this telescope will be five times the size of the largest one in the world."

Perhaps Miller's editor really had sent him to get Einstein's comment on the telescope project, but the question was so weak that it sounds like a desperate stab at locking his foot in the door.

Einstein did not think much of the question. The reporter seemed not to know that visual "proof" of relativity had been announced almost exactly 10 years earlier, or that there was already a 100-inch telescope on Mount Wilson, so the new one would double, not quintuple, the current giant. Einstein made a feeble attempt to brush the question aside with a *mot*, "Not the eye, but the spirit furnishes proof of theories, and that errs most of the time."

Miller noted the comment, marvelously unintelligible, just the sort of impenetrable wisdom you would expect from the world's smartest man. "Is it your opinion," the reporter persisted, "that visual proof of relativity will be disclosed by this or any future telescope?"

Einstein visibly relaxed himself, smoothing his famous hair. The reporter seemed not to realize that bursting in on Einstein while he was in the midst of fruitful thought was like waking somebody from a deeply satisfying dream and pummeling him with daytime banalities. Einstein got a hold on himself. The quickest way to get rid of such people is to answer their immediate questions and discourage others. "The only means of proving relativity with a telescope," he explained in plain language, "is by measuring the deflection of light through the field of gravity. This proof has already been furnished and is correct without the slightest doubt. What one may expect from the great telescope of dimensions not yet constructed lies in another territory— namely exploration of the systems of fixed stars."

Then something happened. Einstein made a mistake. Miller's report does not make clear exactly how it happened. Einstein wanted to send the reporter away.

Miller wanted to stay.

Einstein had important work to get back to.

Miller wondered what Einstein was working on.

Einstein did not wish to say, but he said something that allowed Miller to close his story with the news, "Dr. Einstein then said he was treading on the edge of a great scientific discovery, one that will startle the world far more than relativity."

The next day in New York, the paper's front page carried the headline, "Einstein on Verge of Great Discovery; Resents Intrusion." Soon many of Berlin's newshounds were barking after a story. Einstein quickly realized he had erred in talking to the reporter, but there was no turning off the machinery he had jarred into motion. The streets around the Einstein apartment began to fill with the commotion of flashbulb monkeys, to use Einstein's term for the proto-paparazzi who regularly pestered him.

31

The Saddest Chapter

P rofessional physicists, especially quantum physicists, were not much interested in the rumored coming of Einstein's new field theory. They had made a decisive turn shortly after the Solvay Conference. On January 2, 1928, Paul Dirac submitted a paper to the Royal Society which, in the words of a Danish historian of science, "marked the end of the pioneering and heroic era of quantum mechanics." Dirac's paper managed what Schrödinger had failed to do two years earlier, link special relativity and quantum waves.

When Schrödinger first tried to include Einstein's relativistic mechanics in his wave equation, he did not know about electron spin. Although the spinning electron still had never been physically observed, its effects had been measured. A spinning electric charge produces a magnetic field. Electrons carry a charge and generate a magnetic field; hence, electrons must be spinning. Using this extra knowledge, Dirac had created a miraculously exact rule for calculating the electrodynamics of the atom.

The Dirac equation (actually a set of four equations compressed into one complex expression) can be used to compute the magnetism of electrons. The answer it gives is precisely the one that experimentalists find in the lab. This discovery was the quantum theory's equivalent of Einstein's success at computing Mercury's orbit. Classical notions had failed to predict the observed ratio between the electron's spin rate and its magnetic strength, so when Dirac's equation calculated perfectly the experimentally observed values, physicists had no choice but to nod respectfully.

Dirac's equation could also give the correct values in situations where Schrödinger's equation did work. The Bose-Einstein statistics, too, were folded into the Dirac equation, along with Planck's original quantum equation and Einstein's $E = h\nu$. Even Einstein's most famous baby, $E = mc^2$, was built into the Dirac equation. No wonder Dirac boasted that his theory described "most of physics and all of chemistry."

But talk of Dirac's theory can be misleading to people who take a theory to include some explanation of what is going on. Dirac was not an explainer, neither in physics nor in everyday life. As a lecturer, when asked to explain an idea he would often merely repeat word for word what he had already said. His equation was his theory. Of the Einstein-Bohr debate at Solvay 1927, Dirac said, "I was not very much interested. I was more interested in getting the correct equations. It seemed to me that the foundation of the work of a mathematical physicist is to get the correct equations, that the interpretation of these equations was only of secondary importance." Neither Bohr nor Einstein agreed with Dirac on this point. Perhaps the tip-off is that Dirac called himself a "mathematical physicist" while Bohr and Einstein called themselves theoretical physicists.

The biggest interpretive puzzle was over what Dirac's equation described. It had taken over the ψ of Schrödinger's equation, but what was that squiggle all about? Dirac accepted straight away Max Born's interpretation of it as a statement of probability. Thus, for all Dirac's brilliance in grasping the previous three decades of struggle with the quantum, his equation did not resolve the matters that troubled Einstein. Twenty-three years earlier Einstein had shown that Ludwig Boltzmann's statistical abstraction could be used to measure the real actions of real atoms. He still wanted some real meaning for Schrödinger's and Dirac's mysterious psi. Einstein granted that the equation was "the most logically perfect presentation" of quantum mechanics yet found, but not that it got us any closer to the "secret of the Old One." It neither described the real world phenomena that

Einstein wanted to understand nor proposed new concepts that would make the real world accessible to understanding. Furthermore, Dirac's unification of quantum mechanics with special relativity left out Einstein's later success with general relativity and the gravitational field. (That omission was why the equation covered only "most" of physics.) So Einstein continued pursuing his interest in unifying gravity with electromagnetism.

Berlin's reporters were yammering excitedly about Einstein's anticipated breakthrough, yet they had completely missed the news about how physics had taken a decisive turn. The close of the quantum revolution's "pioneering and heroic era" brought with it a great changing of the guard in physics leadership. The world of commerce has found that it commonly takes one kind of person to found a great enterprise and another kind to direct it in day-to-day competition. In physics, the people who had invented quantum's rules turned out not to be those who were best at using them. Einstein's resistance to post-revolutionary quantum physics is the most famous example, but it was typical of the scientists in this history that they moved in directions away from quantum mechanics. In the summer of 1928, Dirac gave a lecture in Germany that went badly because it predicted negative energy and that seemed a plain impossibility. In this confused state, Heisenberg told Pauli, "the saddest chapter in modern physics is and remains the Dirac theory." Pauli despaired as well and announced that he was abandoning quantum physics. Four years earlier he had moaned he should quite physics and become a film comedian. This time he really did stop doing quantum work and turned to writing a utopian novel. Heisenberg did not retreat so far, but he told Pauli he was forgetting "the more important problems" and returning to pre-Dirac quantum mechanics "in order not to frustrate myself continuously."

Unified physics, too, is often portrayed as Einstein's lonely, futile dream, but others also saw its appeal in one form or another. In early 1928, before turning to his novel, Pauli told Dirac that only a unified field theory would end all the confusion and he urged Dirac to try his hand at finding one. Many years later, Heisenberg, too, began looking for his own unified theory.

The press corps understood very little of this dispute and much of its noise over an idea that even Einstein soon abandoned seems absurd. There were, to be sure, mechanical forces behind this newspaper farce that had nothing to do with Einstein, or science, or the public's eagerness to know. Although by then fame had been industrialized, the supply sources remained preindustrial. There was not yet a cooperating industry for stamping out people eager to be famous. Hollywood was only just beginning to organize itself into the symbiotic relationship that now exists between a media desperate for celebrities and entertainers desperate for shelf space. Without that industrialized supply, the newspapers of 1929 had to cling to the relatively few accidental celebrities and squeeze them for all their juice. At the same time that the papers were besieging Einstein, they were also chasing after Charles Lindbergh and Lawrence of Arabia.

"My name is Mr. Smith," the legendary Lawrence told waiting newsmen when his train pulled into London's Paddington Station. He jumped into a cab, and even though the World War was now 10 years gone, the reporters hopped into other taxis and sped after the old war hero in pursuit of a paragraph.

Undoubtedly, there was much of that sort of nonsense in the roar over Einstein's new work, but Einstein's fame had not been created by the money-craving forces that made, say, Greta Garbo famous. When a group of reporters finally chased him to ground, Einstein pleaded, "I really don't need any publicity." True, but beside the point. The reporters were not trying to do their prey any service. Einstein had stayed famous because the public responded with continuing interest to the story that there was a genius out there for whom the world made sense—more, for whom the *universe* made sense. In this reaction, the trauma of the World War cannot be overstated. It had revealed that a world seemingly based on reason had actually rested on bloody madness. As one popular German antirationalist put it, because of the war "one has had to accept so much that was unimagined, bear such gross things patiently that even now the indignation that one strives to summon up lacks the fitting energy." But there was Einstein, smarter than anybody, and he had found that the world still makes sense. Quantum mechanics, which the press ignored, was what people feared the world had become: arbitrary, technical, and random.

The clue to the hubbub's underlying concern lay in the way the reporters and readers wanted to know what this new theory meant. There was none of today's curiosity about Einstein's personal life. Did he wear boxers or briefs? Who was the lucky woman married to him? There was so little of that side of the story that whenever a report did mention Elsa it smudged her into the background as "Frau Einstein," with no first name given. Instead the justification for the fuss was cited as "the riddle of the universe." Einstein was said to have solved it. When he was in a mood to be generous Einstein recognized a good side to this commotion. He once told a Dutch reporter, "The contrast between the popular estimate of my powers and achievements and the reality is simply grotesque. The awareness of this strange state of affairs would be unbearable but for one pleasing consolation: it . . . proves that knowledge and justice are ranked above wealth and power by a large section of the human race."

Physicists had grown used to seeing a fantastic clamor over Einstein, but this new uproar looked especially scandalous. Bohr especially believed that Einstein was scouting an empty trail, for Einstein was building on general relativity rather than on quantum mechanics. In the British journal Nature, when Bohr finally published his notion of complementarity, he wrote, "general relativity has not fulfilled expectations. A satisfactory solution . . . would seem to be possible only by means of a rational quantum-theoretical transcription of the general field theory, in which the ultimate quantum of electricity has found its natural position as an expression of the feature of individuality characterising the quantum theory."

Whenever a quotation from Bohr runs beyond a clause, it demands a second reading. "Were you never taught in school," Dirac once asked him, "that before you begin a sentence you should have some plan as to how you are going to finish it?" Bohr's general drift, however, was decipherable. It order to "fulfill expectations" general relativity needs a "transcription" into "quantum-theoretical" terms that would unite "the ultimate quantum of electricity" with "an expression" that would identify the ultimate quantum's "natural position." Put more bluntly still: Instead of toying around with the Riemannian geometry that underlies general relativity, Einstein should look squarely to quantum mechanics.

On January 10, 1929, the fame engine moved into even higher gear when Einstein again dispatched Max Planck to the Prussian Academy to present another paper. Publicly the reason for Planck's going in Einstein's stead was Einstein's health, but his health had recovered enough for him to have done the thinking behind the paper. He was, however, not up to fencing with the press, which was in ultrahowl. They were not silenced by Planck's presentation. The paper was short—five pages—but its ideas and mathematics were far beyond the understanding of the unschooled reporters, and they pressed their demand to know what the document meant.

Einstein lay low, but the following day he issued a short statement, "A few days ago [*sic*] I submitted to the Academy of Sciences a work that treats with a novel development of the theory of relativity. The purpose of this work is to unite the laws of the field of gravitation and electromagnetism under a uniform viewpoint." The press dutifully published the paragraph and padded it out as best it could, but two sentences no more satisfied the fame machine than two licks of an ice-cream cone satisfies an unhappy child. Their appetite was only sharpened. A week later the *New York Times* still had nothing to report, so it reported that ignorance—"Amazed At Stir Over New Work, Einstein Holds 100 Journalists At Bay For Week." Meanwhile it kept looking for something to say.

At the end of January, the Prussian Academy published its *Proceedings* containing the papers read that month before the assembly. Normally this journal went straight into obscure libraries. There had been no clamor to see the *Proceedings* back in November 1915 when Einstein announced his general theory of relativity, but in 1929, all the press wanted copies of the January issue. The normal printing of 1,000 copies sold out at once (unheard of!) and 2,000 more were promptly published and seized. Reporters cabled the article. The *New York Herald Tribune* published a translation, along with the equations, the next day. Arthur Eddington wrote Einstein in early February that a London department store had posted the field theory paper in its window and he had seen a mob of ordinary people pushing close to study the thing.

The press was not yet satisfied and still demanded: What does it mean? The *New York Times* and the London *Times* managed together to persuade Einstein to write an explanation of his theory in "simplified" terms. Before turning it over to a reporter for cabling to New York, Einstein had Elsa read it back to him. When she finished Einstein said, "The article has permanent value and I shall use it in a book later." He never did because he soon abandoned the "distant parallelism" that underlay the theory's mathematics. It is too bad that the article was not reprinted and curious readers are still forced to suffer through scratched microfilm to find it. The article provides more than a discussion of field theory. Most of it presents an excellent summary by Einstein himself of the Newtonian view he overturned and how he did it. Even the article's latter part spent as much space explaining what Einstein was trying to do as it did going into details about what he had done.

"The characteristics which especially distinguish the general theory of relativity and even more the new third stage of the theory, the unitary field theory, from other physical theories are," Einstein reported bluntly, "the degree of formal speculation, the slender empirical basis, the boldness in theoretical construction and, finally, the fundamental reliance on the uniformity of the secrets of natural law and their accessibility to the speculative intellect." Einstein summed himself up there about as well as it can be done. The difference between relativity and other physical laws was not in its notions of time, space, light, or geometry. It was speculation, boldness, and a reliance on the uniformity and accessibility of natural law. And it was that faith in accessible, natural law that made all the noise over Einstein fundamentally unlike the crowds swooning over one more pop personality. People wanted to believe that boldness, brilliance, and natural law still worked.

Einstein went on to mention that there were physicists who disliked his approach although he did not mention Mach, Bohr, or anybody else by name. Instead, he cited a philosopher he did agree with: Emile Meyerson, whose 1908 book, *Identity and Reality*, argued that scientific knowledge tries to get beyond mere description and predictive laws to an understanding of the reality beyond the appearances.

And once it got that explanation the press calmed down. Einstein slipped off to Berlin's Wannsee suburb for an undisturbed rest. The pity and foolishness of the commotion is that in chasing Einstein down and getting him to spell out what he was up to, it missed the bigger story. The press took for granted that law both underlies and guides one to reality, and it ignored the ongoing quarrel between Einstein and Bohr.

A closer comparison of their comments on what a unified theory must do gets at the heart of the difference between them. In his newspaper article Einstein, like Bohr in his complementarity essay, said he was looking for a "mathematical expression." Bohr, however, wanted an expression "characterizing the quantum theory" while Einstein wanted "the mathematical expression of the physical fields." The difference? Bohr's math characterized theory while Einstein's looked to physical reality.

This subtle divide was behind the Einstein-Bohr quarrels from the beginning. Untangling its depths can lead to arcane distinctions, but these slippery differences had led to a decade of dispute. In practice, it had come down to where, when you must, you allow your ambiguities and contradictions. In the end, of course, both wanted a coherent, unambiguous physics, but while the search persisted, Bohr preferred an ambiguous reality to a contradictory theory. He was like a novelist who tells a coherent story by finding a symbol that captures the ambiguity of an experience. He had fiercely resisted the way Einstein's photon proposal confused the distinction between waves and particles. Meanwhile, Einstein saw contradictions between ideas as the key to finding an unambiguous, coherent reality. Of course he balked at the way quantum mechanics used formal symbols instead of the names of real things.

32

A Reality
Independent of Man

O n May Day, 1929, Berlin's communists began their annual
parade through the city's working class district. As they moved
along, chanting slogans and singing the International, the police began forming tactical squads. The chanting and marching continued. It was a traditional celebration and, that year, it was also an
open resistance to the city's socialist police chief who had banned all
public demonstrations. The line of marchers stretched for blocks when
a wing of police moved in to bar their advance. Stopped in front, the
demonstrators felt the pressure of their comrades behind, coming in
from the rear, unclear about what exactly was happening. What did
happen next is lost in the fog of contradictory testimony. Maybe somebody threw something. At some point the police definitely opened
fire. A photograph of the event tells nothing—people are running in
every available direction. In the end, hundreds were wounded; approximately 30 were dead.

Berlin was becoming a poor place to live. The murderous style of
politics had returned. Beatings were popular. Lies were doing battle
with rival lies. Killings happened as well. Two political parties had
organized as military forces. Those quick-stepping, head-bashing types
who had carried Mussolini to Italy's top had became rampant in Berlin. They pretended to be communists or nationalists or whatever else
would support their thug instincts. The capital had quickly changed

into an arena for the young, the brutal, and the stupid. After Moscow, Berlin had the largest communist apparatus in the world and it had no trouble organizing a *Rote Frontkämpferbund* (red fighter group) of hooligans. Meanwhile, Goebbels had shown himself woefully capable of conjuring hatred from unpromising soil. Young Nazi thugs were proving as energetically mean and stupid as the red fighter membership. Men with hates organized them into automatons of action, loathers of the strain that imagination requires. In another few years, robots would be tossing books into a pyre before Berlin's opera house, cheering and celebrating their contempt for ideas. A decade after the Treaty of Versailles, Germany had recovered its status as one of the leaders of European civilization. Its physics and mathematics along with its literature, drama, music, and philosophy would influence the world throughout the coming century. But its future in Germany itself looked bleak. The same thing had happened elsewhere. Italy lost its position in the vanguard of natural philosophy after the Catholic Church condemned Galileo to silence, and after the French guillotined Lavoisier they watched their country surrender its lead in chemistry. But the rampaging thugs in Berlin's streets were proudly ignorant of European history and, therefore, could not take its warning.

Einstein prudently chose this time to buy a summer home in Berlin's southwest countryside, at a village called Caputh. The house they chose (Elsa, of course, was the one who found it) was far from attractive. It had an ugly upstairs balcony that dominated the main structure, but it was practical. Caputh was surrounded by waterways where Einstein could satisfy his love of sailing. The balcony provided a relaxing space for a small social ensemble. More importantly, Caputh was remote. Reporters looking for a quick story could not get out there easily, and the house did not even have a telephone. To reach the village, visitors rode a train from Berlin to Potsdam and then transferred to a bus at the Potsdam station for the six mile jaunt to Caputh. Baedeker's guide reported that when the bus was not available, travelers could walk from Potsdam to Caputh in two hours.

It was a grand shelter from Berlin's sudden angers. Germany had seemed to recover from both the war and the postwar. Then the American stock market failed; foreign investment in Germany halted.

Worse, foreign investors also began calling in outstanding loans, toppling many businesses into bankruptcy. Jobs evaporated. So soon after the epic inflation, people had few savings to fall back on. By December, only two months after the crash, Germany's Secretary of State reported that Berlin had returned to horrors not seen since the "terrible years of 1918-19." But Einstein was safely out of the way of the gangs and in no danger of unemployment. He was still one of the people whom Berlin's most distinguished visitors hoped to meet.

One such traveler was "Rabbi" Tagore. He returned to Germany in the summer of 1930 and made the trek to Caputh to pay his respects to the famous scientist. Einstein had been forewarned and came down the road to meet his guest. He escorted Tagore and the rest of his entourage up the hill to his home, and then up the stairs to the balcony where they could drink tea and converse. Welcoming though he was, the situation was awkward because of language problems. Tagore spoke no German and Einstein's conversational English was still weak.

Tagore's group included two women, thought to have been his sister and her daughter. Both wore saris of many colors. Doubtless they looked splendid as they came up the hill to the summer house, and the dull birds in the Einstein party highlighted the Indian colors even more. Besides Elsa, there was her youngest daughter Margot and Margot's husband. A photo of the visitors and their hosts shows that the westerners were dressed in shades of black with little touches of white.

Tagore's secretary was also on hand. He translated and took notes. Margot's husband took notes as well and, to Einstein's displeasure, a few weeks later published an account of the meeting in the *New York Times*. The transcript, Einstein felt, should never have been published, but inevitably it was. Who could resist reporting a meeting of East and West? Einstein enacted his natural role as defender of the West's central "Faustian" belief that fixed laws define the reality beyond us, while Tagore played the mystical, "oriental" guru who turned inward for his truth.

The transcripts produced during the conversation were short on "please pass the cookies" type moments, and they read something like a radio interview, but for all his differences with Tagore, Einstein appears to have found the exchanges agreeable. A month later, when Tagore returned to Berlin from a tour of the German hinterlands, Einstein took the bus and train into the capital to meet him again.

The heart of the hilltop conversation that captured the contrast between the two began when Einstein asked, "If there would be no human beings any more, the Apollo of Belvedere would no longer be beautiful?" The record does not show whether the translator needed to elaborate on what the Apollo of Belvedere was.

"No!" Tagore said and Einstein needed no translator.

"I agree with regard to this conception of Beauty, but not with regard to Truth."

"Why not?" Tagore asked. "Truth is realized through man," giving the answer Bohr might have given had he been prone to making clear statements.

"I cannot prove that my conception is right, but that is my religion." Einstein's talk of religion was probably a bow toward Tagore. In answering questions from a Japanese scholar a year earlier he had said that the term "'religious truth' conveys nothing clear to me at all."

Tagore, according to the *Times* story, spoke with "majestic tranquility, as if reciting a poem or a sermon." He answered Einstein, "Beauty is in the idea of perfect harmony which is in the Universal Being. Truth is the perfect comprehension of the Universal Mind. We individuals approach it through our accumulated experience, through our illumined consciousness—how otherwise, can we know Truth?"

Einstein might have said he recognized Truth through the authority of tested experience, the logic of explanation, and the understanding that the two together provide. Instead, he adopted a humble tone. "I cannot prove scientifically that Truth must be conceived as a Truth that is valid independent of humanity; but I believe it firmly. I believe, for instance, that the Pythagorean theorem in geometry states something that is approximately true, independent of the existence of Man. Anyway, if there is a reality independent of Man there is also a Truth relative to this reality; and in the same way the negation of the first endangers a negation of the existence of the latter."

Tagore disagreed, "Truth, which is one with the Universal Being, must essentially be human, otherwise whatever we individuals realize as true can never be called Truth—at least the Truth which is described as scientific and can only be reached through the process of logic, in other words, by an organ of thoughts which is human."

A startling feature of this exchange is that the parts that are comprehensible to western ears parallel the issues in the ongoing debate Einstein was having with Niels Bohr. Particularly notable was Einstein's coda to his claim that if there is a reality beyond humans, there is also a truth that is independent of us: "The negation of the first [a reality independent of humanity] endangers a negation of the existence of the latter [truth]." Why add this business, which seems self-evident after the first part? Perhaps Einstein was spelling out to himself what was at stake in the dispute with Bohr. If you cannot find reality, how can you find truth? And if you cannot find truth, what claims can you make for science?

"The problem begins," Einstein said a few minutes later, "[with] whether Truth is independent of our consciousness."

"What we call truth," Tagore insisted, "lies in the rational harmony between the subjective and objective aspects of reality, both of which belong to the super-personal man."

It is notable that Einstein never probed Tagore's references to "the super-personal man." That was clearly an Indian doctrine, a novelty for Einstein, a different way of thinking about the "Old One" whose secret Einstein chased. However, he seems to have had no interest—not even the superficial interest of the generous host—in Tagore's religious ideas; indeed he expressed the depth of his separation from Tagore a few minutes later when he blurted in wonderment, "Then I am more religious than you are."

He should have said, "Then I am more *western* than you are," because he was reacting in a witty, professorial way to Tagore's statement, "If there be some Truth which has no sensuous or rational relation to the human mind, it will ever remain as nothing so long as we remain human beings."

Einstein, and Bohr, too, had been reared in the western one-God-of-Abraham tradition in which the true, real, eternal is out there and we are nothing compared to it. Even western atheism accepts that premise, merely rejecting the consolation of a God who oversees creation and is concerned for the anguish of our lives. Both Einstein and Bohr had devoted their energies to studying what lay outside themselves, and neither of them knew what to think of Tagore's definition of his own, Indian, religiousness: "My religion is the reconciliation of the Super-personal man, the Universal human spirit, in my own individual being."

33

A Certain Unreasonableness

To months later, on September 15, 1930, the Nazis astonished themselves as well as the world by winning the second largest number of seats in the Reichstag. Hitler had found that power in the modern world comes only secondarily from the barrel of a gun. Foremost it comes from lies. The authority of Nazi lies rested on personal appeal and loyalty. Their explanations cast blame. They replaced understanding with excitement. And there was plenty of excitement in early October, when the Reichstag opened with its new membership. Demonstrations took place across Berlin. Count Kessler saw them in the part of town where Einstein caught his train to Potsdam. Kessler pronounced the demonstrators mostly "adolescent riff-raff" and watched them shout, "Germany awake! Death to Judah." Hot-blooded ignorance was on the march.

A week after this terrifying scene, Einstein was back in Brussels for the opening of another Solvay Conference. This time, of course, there had been no controversy about inviting the Germans. Einstein came as a matter of routine. With Lorentz dead, the gathering had a new chairman, Paul Langevin, but the idea was the same: bring the world's greatest physicists into a room and start talking. The official topic that year was "Magnetism." Bohr came, looking (according to the photographs) much older than he had 10 years earlier when he first met Einstein. It was now six years since Fritz Haber had spoken about how these "two . . . stand in such deep opposition." The description still held. These two men still seemed to grasp the stakes far more deeply than their colleagues: Can science claim an objective knowledge of how nature governs itself or boast only of a practical understanding of what experiments will show? Bohr might have thought the matter settled, but Einstein had been saving something for him. The two men sat together; a photograph shows them sunk into easy chairs smoking pipes and talking.

Einstein proposed one of his elaborate contraptions for getting around Heisenberg's principle, in this case the part about not being able to know both time and energy with great certainty. You could, Einstein proposed, create a hollowed ball and heat it up so that the interior would fill up with radiation. This was exactly the sort of experiment that Planck had considered 30 years before when he introduced the quantum h whose tortuous path had led them to their quarrel. In those early days Planck's "blackbodies" had included a small hole through which radiation escaped for study. Einstein proposed that this hole be covered by a shuttered gate that moved so quickly it could release its radiation one photon at a time.

Bohr listened alertly. Give a point to Einstein there for having discovered photons in the first place.

Now attach a clock to the shutter so that we can tell the exact time it opens and a photon escapes. This system allows for a very precise knowledge of the time. That takes care of the time side of the uncertainty principle, but what about measuring the energy?

Here Einstein resorted to another of his famous discoveries, $E = mc^2$. The photon that escapes is pure energy, as the long agony over the photon's $h\nu$ had finally established. Any attempt to measure the photon's energy directly would be subject to a variety of limits that would seem to validate Heisenberg. Einstein proposed an indirect measure. Weigh the blackbody before opening the shutter and then again after closing the shutter. This weighing, which can be extremely accurate, will show a drop in mass. Just as the sun loses mass every instant that it radiates energy into the universe, Einstein's blackbody will lose mass as it radiates its little photon.

Sure, sure, the drop will be tiny, way too small for any real scale to detect, but we are imagining an experiment here. The body loses mass and there is no principled reason that a scale could not measure it. Then, once we know the loss in mass, we can use the $E = mc^2$ equation to compute how much energy that mass represents. We will have measured the time and energy of the photon with much greater accuracy than Heisenberg's principle allows.

Bohr was thunderstruck. The room of physicists was stunned. Einstein had done it, gotten around Heisenberg, Bohr, and all those pragmatists who thought they had chased reality's claims from the heart of physics. There was, after all, more in heaven and earth than in their equations.

Léon Rosenfeld, who later became Bohr's longtime collaborator, was at Solvay 1930 and reported that, "During the whole of the evening [Bohr] was extremely unhappy, going from one to the other, and trying to persuade them that it couldn't be true, that it would be the end of physics if Einstein were right, but he couldn't produce any refutation."

Bohr might have thought this experiment destroyed physics, but Einstein felt he had saved it by showing there was still more work to do. When he began his life as a physicist, the leaders were confident that Newton's mechanics, Maxwell's electromagnetism, and the new laws of thermodynamics summed up the whole of nature. By picking away at loose threads and examining areas that others had told him were pointless to question, Einstein had shown that more could be done. Through it all, Einstein had maintained a faith in God's honesty. In his first theory of relativity he had made that faith a principle: The laws of nature are the same everywhere in the universe. This kind of principle can never be proved, but Einstein built on it, confident that "the Lord is . . . not malicious." With quantum theory he again refused to believe that reality was permanently hidden.

The next morning, however, Bohr reported that he had more to say. It is true, Bohr agreed, that we can imagine a device that is so sensitive it can weigh the lost photon, but the measuring device will still have to be a scale of some kind. There are many different kinds of scales—doctors' scales, grocers' balances, and so on. Any system will do, but let us imagine the simplest one, a one-sided balance that points to a chart. The balance moves up and down, and as it does, a pointer moves with it up and down across a scale. Because this balance moves to different positions for different weights it will move—very slightly, true, but move—when the blackbody loses its photon.

Where will it move? To its new position, yes, but as the photon departs, its momentum will push the balance in the opposite direction. In short, as the balance moves to its new position, it will sway a bit, as balances always do when you change the weights involved. We cannot predict exactly how the balance will sway because we cannot predict the direction taken by the escaping photon, but we know it will rock gently like a suitcase tag swaying on a train.

Perhaps Einstein's famous hair stood even more starkly on end as Bohr managed to slip an uncertainty into the system. Einstein had always been troubled by his inability to calculate exactly what direction an emitted photon will take.

Now, Bohr continued, it is while this balance sways in its unpredictable path that the clock attached to it will measure the time. But we know from general relativity that the time on a moving clock differs from the time of observers. Of course the equations of general relativity would let us translate that clock's time to our time if we knew the clock's speed and direction, but in this case we do not know how the clock is swaying, so we cannot calculate the time difference exactly. Thus, although we can know the energy to great precision, there is still an uncertainty about the time. Heisenberg's principle holds.

Bohr might as well have slapped Einstein with a fish. The founder of relativity had forgotten that time is not absolute. Still, even in this defeat, Einstein remained a *mensch* and rose to the moment. He was much the better mathematician and helped Bohr work through the calculations to show that the uncertainty between time and energy would, after all, be as Heisenberg's principle said. Quantum mechanics continued to describe experimental outcomes without revealing an underlying reality.

Einstein's inability to find inherent contradictions in quantum mechanics appears finally to have cracked the optimism that had been fundamental to his scientific character. After that morning he was no longer sure he had the imagination to find the secret behind the compass. He had told Count Kessler that doing profound physics was really very simple, you looked for the point where physics contradicted itself and found a way to resolve the contradiction. When he said that, he had not doubted his talent for solving the puzzles. Now the contradictions in quantum physics seemed to be right on the table—if you do one set of experiments you can show that light and electrons are waves; if you do another set of experiments, you show these very same things are particles—yet he could not reduce the paradox to one experiment that brought the contradictory measurements together.

He believed that quantum mechanics was both right and a dead-end, providing no pathway to the rest of physics. For the rest of his professional life he looked for a unified theory that would provide an entryway into a coherent, *realistic* quantum physics. He did not succeed, and few colleagues joined him in the search. For them, Bohr had won, scientific realism had fallen and with it the belief that the world is objectively comprehensible.

Many physicists said that what had become of Einstein was a great tragedy. Physics had to do without him. Einstein never saw himself as tragic, nor his defiance as futile. His state of mind was, as he once said about Max Planck, "akin to that of the religious worshipper or the lover; the daily effort comes from no deliberate intention or program but straight from the heart." When those late Renaissance dreamers imagined that the natural world could be explained in natural terms, many had opposed their idea as blasphemy and presumption that was dangerously apt to lead one away from God and toward materialism. The critics had a point. The search for natural laws proved so successful that for many savants the search became an end in itself. They did not follow the Renaissance expectation that discovering order would bring them closer to God. Einstein's talk of using physics to find the secret of the Old One sounded poetic and eccentric. With the triumph of quantum mechanics and Bohr's complementarity, physicists put away all pretense of striding closer to knowing what was out there beyond ourselves.

For the professional physicist this change was of minor interest, and you sometimes read a professional who denies that the Einstein-

Bohr debates were of much importance. The difference between logical laws that work because they are efficient and natural laws that work because they get the meaning right is of no interest to their pragmatic labors. Quantum mathematics provided a way to test predictions and organize the complete results of any experiment. It was the amateurs who, without knowing clearly what had happened, reacted to the change in physics. Quantum physics had become permanently paradoxical and alien, a set of techniques that lay people could read about over and over and yet never seem to really understand. Like the most advanced music, painting, and literature of the period, physics took a turn that its old audience could neither follow nor grow into. The subject could not be dismissed; after all, physicists had discovered the secret of blowing up the world, but it seemed to say nothing meaningful about either their lives or their world.

Einstein refused to follow his colleagues down that pathway. A few weeks after Bohr's refutation at Solvay, Einstein was with his friend Paul Ehrenfest in the Netherlands, participating in a panel discussion on quantum physics. As Einstein criticized quantum mechanics, some of the distinguished scientists grew quietly restless. A speaker on the floor then defended quantum theory as complete and consistent. Einstein was not happy. "I know this business is free of contradictions," he granted but waved triumphant opinion aside as he had done a dozen years earlier in the Reichstag, "yet in my view it contains a certain unreasonableness." The satisfied revolutionaries in the room could only stare. Their pathfinder was defying them all, determined still to find the enduring face behind all our changing facts.

Afterword

The 1930 Solvay Conference was the last Solvay Einstein attended. The next one took place in 1933, and by then he was in the United States, a refugee from the Nazi government that was voted into office in March 1933. Einstein lived 25 years beyond Solvay 1930, Bohr more than 30. They had one more long-distance dispute. In 1935, Einstein published an article pointing to a problem he saw entangling quantum theory, but Bohr answered him sharply and for many decades was considered triumphant. Only in the twentieth century's final quarter did physicists realize that Einstein had been back in his crow's nest spotting still more unanticipated trouble from quanta. Most importantly, the great equations of the quantum revolution never failed. Schrödinger's equation, Dirac's equation, and Heisenberg's principle have all withstood the tests of decades. Nothing hidden has been found, nothing extra has been needed, nothing exceptional has been acknowledged. The accuracy of the equations has survived the ever-improving precision of measurements, and nobody expects them to falter. Even the puzzle of negative energy in Dirac's equation proved correct when "antimatter" was discovered.

Bohr and Einstein never met again in Europe but did meet several more times in America. After 1930, however, matters between them were never the same. Bohr's idea of complementarity and his Copenhagen Interpretation enjoyed a monopoly over quantum physics for another generation. In the 1950s, however, the kind of quarrel-ing over physical meaning that had seemed to end with Newton re-turned when a series of rival interpretations of the same logical laws started to appear. Perhaps the most famous, or notorious, of these is the "many worlds interpretation" in which the universe keeps splitting into as many alternate realities as are required to produce every pos-sible outcome for each quantum event. The "sum over histories" inter-pretation proposes that photons, electrons, and other quantum entities follow all their possible trajectories at the same time. Meanwhile, "hid-den variable" interpretations propose that reality still determines what happens as firmly as it did during the reign of Newton, but the details of these determinants are hidden from us. This last interpretation builds on ideas Louis de Broglie had proposed at the start of the quantum revolution.

Of course, there were some proud moments yet to come for the quantum revolutionaries. De Broglie had won the 1929 Nobel Prize. Heisenberg won it in 1932. Schrödinger and Dirac shared 1933's prize. Pauli finally landed his in 1945. Max Born did so much for the quan-tum revolution yet was somehow overlooked by the Swedes until 1954.

Despite these honors, our history's final scene in Holland smells of doom. Ehrenfest was rocked with despair and would, in three years, kill his son, Vassik, and then himself. The other professors in the room would be scattered to the earth's corners. The Jews among them would have no choice but to flee. Others, like Schrödinger, could have stayed and been honored, but they fled as well. The heritage of Göttingen, preserved by Napoleon, would quickly be destroyed by the Nazis as its brain power was chased abroad. German nationalists had done what Planck had said no enemy could do, rob German science of its posi-tion in the world. The forgotten figures of Philipp Lenard and Johannes Stark would suddenly move to the top of Germany's scien-tific bureaucracy.

Elsa died in 1936 while Einstein looked on in helpless misery. "I never thought he was so attached to me," Elsa wrote a friend in amaze-ment over her husband's worry, "That, too, helps." But Einstein did not live alone for long. His secretary, Helen Dukas, remained with him and he brought some of his family over. Matters were growing more terrible in Europe every day. He got his oldest son, Hans Albert, and his family out in 1937. His sister, Maja, who had not played much part in the Einstein story since his boyhood, joined him in Princeton in 1939 and stayed until her own death in 1951. During all his years in America Einstein pursued both his physics and his humanist politics. His anti-fascism, pacifism, enthusiasm for democracy, and support for civil rights brought him to the attention of America's policeman-in-chief, J. Edgar Hoover. The FBI complied a fat dossier on him but never acted on it. As had been the case 30 years earlier in Germany, America's leading politicians thought he was a malodorous flower, but, after he became a U.S. citizen in 1941, they proudly showed him off in their buttonholes.

Bibliography

A. WHERE TO BEGIN

Bernstein, Jeremy. *Albert Einstein and the Frontiers of Physics*. New York: Oxford University Press, 1996. [A short overview of Einstein's life and physics.]

Born, Max. *Born-Einstein Letters*. Translated by Irene Born. Commentary by Max Born. New York: Walker Publishing, 1971. [Readable schmoozing between two greats.]

Calaprice, Alice, ed. *The Expanded Quotable Einstein*. Princeton: Princeton University Press, 2000. [A surprisingly vivid presentation of Einstein's breadth and character.]

Cassidy, David C. *Uncertainty: The life and science of Werner Heisenberg*. New York: W.H. Freeman and Company, 1992. [A readable biography.]

Einstein, Albert. "Autobiographical Notes" (1949). In Schlipp (1949), pp. 1-94. [Einstein explains himself in plain language. Also available in a stand-alone edition: *Autobiographical Notes: A Centenniel Edition* La Salle, Ill.: Open Court Publishing, 1991.]

Feynman, Richard. *The Character of Physical Law*. New York: The Modern Library, 1994. [A clear view of physics as it looks after the quantum revolution.]

Fölsing, Albrecht. *Albert Einstein: A biography*. Translated by Ewald Osers. New York: Penguin Books, 1997. [The most up-to-date of the big biographies.]

Frank, Philipp. *Einstein: His life and times* New York: Da Capo Press, 1947. [The classic study of Einstein's character, by one who knew him.]

Frayn, Michael. *Copenhagen*. New York: Anchor, 2000. [Full of light, chatty talk about the quantum revolution.]

Friedrich, Otto. *Before the Deluge: A portrait of Berlin in the 1920s*. New York: Harper Perenniel, 1995. [A portrait of Einstein's Berlin.]

Gay, Peter. *Weimar Culture: The outsider as insider*. New York: W.W. Norton & Company, 2001. [Clear account of Einstein's Germany.]

Gribbin, John. *Q is for Quantum: An encylopedia of particle physics*. New York: Simon & Schuster, 1998. [An excellent reader for delving into any question about quantum mechanics.]

Kessler, Harry. *Berlin in Lights: The diaries of Count Harry Kessler (1918-1937).* Translated and edited by Charles Kessler. New York: Grove Press, 1999. [The most enjoyable on-the-scene sources for Einstein's Berlin.]

Kragh, Helge. *Quantum Generations: A history of physics in the twentieth century.* Princeton: Princeton University Press, 1999. [Fine overview.]

Pais, Abram. *"Subtle is the Lord. . .": The science and life of Albert Einstein.* Oxford: Oxford University Press, 1982. [The best biography of Einstein, even if you skip the math.]

— *Niels Bohr's Times in Physics, Philosophy, and Polity.* Oxford: Clarendon Press, 1991. [The best biography of Bohr. Written with the lay reader in mind.]

B. EINSTEIN AND THE QUANTUM

Regrettably, there are no accessible books that make clear what Einstein's antipathy to quantum mechanics was all about. Bolder readers might want to try:

Beller, Mara. *Quantum Dialogue: The making of a revolution.* Chicago: The University of Chicago Press, 1999. [There is much in this book, but it assumes the readers already know the story.]

Fine, Arthur. *The Shaky Game: Einstein, realism, and the quantum theory.* Second edition. Chicago: The University of Chicago Press, 1996. [This important work also assumes the reader brings much knowledge to the table.]

Whitaker, Andrew. *Einstein, Bohr and the Quantum Dilemma.* Cambridge: Cambridge University Press, 1996. [The more you already know about physics, the more you will get out of this valuable study.]

C. SOURCES: BOOKS

Albert, David Z. *Quantum Mechanics and Experience.* Cambridge, Mass.: Harvard University Press, 1992.

Baedeker, Karl. *Berlin and its Environs: Handbook for travellers.* Leipzig: Karl Baedeker, Publisher, 1923.

Bennett, Arnold. *From the Log of the Velsa.* New York: The Century Co., 1914.

Berlinski, David. *Newton's Gift: How Sir Isaac Newton unlocked the system of the world.* New York: The Free Press, 2000.

Blotner, Joseph. *Faulkner: A biography.* 2 volumes. New York: Random House, 1974.

Bohr, Niels. *Collected Works.* Volume 4: The Periodic System. 1920-1923. Edited by J. Rud Nielsen. Amsterdam: North-Holland Publishing Company, 1977.

Bolles, Edmund Blair. *A Second Way of Knowing: The riddle of human perception.* New York: Prentice-Hall Press, 1991.

Bolles, Edmund Blair, ed. *Galileo's Commandment: An anthology of great science writing.* New York: W.H. Freeman, 1997.

Born, Max. *My Life: Recollections of a Nobel laureate.* New York: Charles Scribner's Sons, 1978.

Brian, Denis. *Einstein: A life.* New York: John Wiley & Sons, 1996.

Brod, Max. *The Redemption of Tycho Brahe.* Translated by Felix Warren Crosse. New York: Alfred Knopf, 1928.

Brown, Melvyn. *Satyen Bose: A life.* Calcutta: Annapurna Publishing House, 1974.

Cézanne, Paul. *Letters.* Edited by John Rewald. Translated by Margarite Kay. Oxford: Bruno Cassirer, 1941.

Chernow, Ron. *The Warburgs: The twentieth-century odyssey of a remarkable Jewish family.* New York: Random House, 1993.

Clark, Ronald W. *Einstein: The life and times.* New York: Avon Books, 1971.

Clegg, Brian. *Light Years and Time Travel: An exploration of mankind's enduring fascination with light.* New York: John Wiley & Sons, 2001.

Crowe, Michael J. *Theories of the World: From antiquity to the Copernican revolution.* Second revised edition. Mineola, New York: Dover Publications, 2001.

Dachy, Marc. *The Dada Movement, 1915-1923.* New York: Rizzoli International Publications, 1990.

Döblin, Alfred. *Berlin Alexanderplatz: The story of Franz Biberkopf.* Translated by Eugene Jolas. New York: The Continuum Publishing Company, 2002.

Drake, Stillman. *Galileo at Work: His scientific biography.* New York: Dover Publications, 1978.

Ehrenfest, Paul and Tatiana Ehrenfest. *The Conceptual Foundations of the Statistical Approach in Mechanics.* New York: Dover Publications, 1990.

Einstein, Albert. *Collected Papers:* Volume 1, "The Early Years, 1879-1902 [ed., John Stachel et al., 1987]; Volume 2, "The Swiss Years: writings 1900-1909" [ed., John Stachel et al., 1989]; Volume 3, "The Swiss Years: writings, 1909-1911." [ed. Martin J. Klein et. al., 1993]; Volume 4, "The Swiss Years: writings, 1912-1914." [ed. Martin J. Klein et al., 1995]; Volume 5, "The Swiss Years: correspondence, 1902–1914." [ed. Martin J. Klein et al., 1993]; Volume 6, "The Berlin Years: writings, 1914-1917." [ed. A.J. Kox et al., 1996]; Volume 7, "The Berlin Years: writings, 1918-1921." [ed. Michel Janssen et al., 2002]; Volume 8A, "The Berlin Years: correspondence, 1914-1917." [ed. R. Schulmann et al., 1997]; Volume 8B, "The Berlin Years: correspondence, 1918." [ed. R. Schulmann et al., 1997]. Princeton: Princeton University Press, 1987 -.

— *H. Lorentz: His creative genius and his personality.* Leiden: National Museum for the History of Science, 1953.

— *Ideas and Opinions.* New York: Three Rivers Press, 1982b.

Einstein, Albert and Mileva Maric. *The Love Letters.* Edited by Jürgen Renn and Robert Schulmann. Princeton: Princeton University Press, 1992.

Ellmann, Richard. *James Joyce.* Oxford: Oxford University Press, 1982.

Elon, Amos. *The Pity of It All: A history of the Jews in Germany, 1743-1933.* New York: Henry Holt and Co., 2002a.

Eyck, Erich. *A History of the Weimar Republic.* Translated by Harlan P. Hanson and Robert G.L. Waite. 2 vols. Cambridge, Mass.: Harvard University Press, 1967.

Farmelo, Graham, ed. *It Must Be Beautiful: The great equations of modern science.* London: Granta Books, 2002.

Fenneberg, Paul. *This Copenhagen: An incomplete introduction to a town I love.* Copenhagen: Scandanavian Publishing Company, 1946.

Feuchtwanger, E.J. *From Weimar to Hitler: Germany, 1918-33*. Second edition. New York: St. Martin's Press, 1995.

Ford, Ford Maddox. *The Good Soldier: A tale of passion*. London: Penguin Books, 1947.

Fry, Joan Mary. *In Downcast Germany: 1918-1933*. London: Clarke, 1944.

Fussell, Paul. *The Great War and Modern Memory*. London: Oxford University Press, 1971.

Galilei, Galileo. *Dialogues Concerning Two New Sciences*. Translated by Henry Crew & Alfonso de Salvio. New York: Dover Publications, 1954.

— *Dialogue Concerning the Two Chief World Systems: Ptolemaic and Copernican*. Translated by Stillman Drake. 2nd revised edition. Berkeley: University of California Press, 1967.

Goddard, Peter, ed. *Paul Dirac: The man and his work*. Cambridge: Cambridge University Press, 1998.

Gullberg, Jan. *Mathematics from the Birth of Numbers*. New York: W.W. Norton, & Company 1997.

Haas-Lorentz, G.L. De, ed. *H.A. Lorentz: Impressions of his life and work*. Translated by Joh. C. Fagginer Auer. Amsterdam: North-Holland Publishing Company, 1957.

Heilbron, J.L. *The Dilemmas of an Upright Man: Max Planck and the fortunes of German science*. With a new afterword. Cambridge, Mass.: Harvard University Press, 2000.

Hemingway, Ernest. *Dateline: Toronto: The complete Toronto Star Dispatches, 1920-1924*. Edited by William White. New York: Charles Scribner's Sons, 1985.

Herodotus. *The Histories*. Translated by Aubrey de Selincourt. London: Penguin Books, 1954.

Highfield, Roger and Paul Carter. *The Private Lives of Albert Einstein*. New York: St. Martin's Press, 1994.

Holton, Gerald and Yehuda Elkana, eds. *Albert Einstein: Historical and cultural perspectives*. The Centenial Symposium in Jerusalem. Mineola, New York: Dover Publications, 1997.

Jammer, Max. *The Conceptual Development of Quantum Mechanics*. Los Angeles: American Institute of Physics, 1989.

Jerome, Fred. *The Einstein File: J. Edgar Hoover's secret war against the world's most famous scientist*. New York: St. Martins Press, 2002.

Kantha, Sachi Sri. *An Einstein Dictionary*. Westport, Conn.: Greenwood Press, 1996.

Klein, Martin J. *Paul Ehrenfest*. Volume 1, "The Making of a Theoretical Physicist." Amsterdam: North Holland Publishing Company, 1970a.

Koestler, Arthur. *The Sleepwalkers: A history of man's changing vision of the universe*. New York: Arkana, 1990.

Kuhn, Thomas S. *The Structure of Scientific Revolutions*. Second edition, enlarged. Chicago: University of Chicago Press, 1970.

— *Black-Body Theory and the Quantum Discontinuity, 1894-1912*. Chicago: Universiy of Chicago Press, 1978.

Kurzke, Herman. *Thomas Mann: Life as a work of art*. Princeton: Princeton University Press, 2002.

Laqueur, Walter. *Weimar: A cultural history 1918-1933*. London: Phoenix Press, 2000.

Leach, Dr. Henry Goddard, ed. *Living Philosophies*. New York: Simon & Schuster, 1931.

Liesner, Thelma. *Economic Statistics: 1900-1983: United Kingdom, United States of American, France, Germany, Italy, Japan.* London: The Economist, 1985.

Mach, Ernst. *The Science of Mechanics, a critical and historical account of its development.* Translated by Thomas J. McCormack. La Salle, Ill.: Open Court Publishing, 1942.

Marage, Pierre and Grégoire Wallenborn, eds. *The Solvay Councils and the Birth of Modern Physics.* Basel: Birkhäuser Verlag, 1999a.

Mehra, Jagdish, ed. *The Physicist's Conception of Nature.* Dordrecht, Holland: D. Reidel Publishing, 1973.

Mehra, Jagdish. *The Solvay Conferences on Physics: Aspects of the development of physics since 1911.* Dordrecht, Holland: D. Reidel Publishing, 1975a.

Meyerson, Emile. *The Relativistic Deduction: Epistemological implications of the theory of relativity.* Boston: D. Reidel Publishing, 1985.

Monk, Ray. *Ludwig Wittgenstein: The duty of genius.* New York: Penguin Books, 1991.

Ord-Hume, Arthur W. G. *Perpetual Motion: the history of an obsession.* New York: St. Martin's Press, 1977.

Overbye, Dennis. *Einstein in Love: A scientific romance.* New York: Viking, 2000.

Payer, Lynn and Kerr L. White. *Medicine and Culture: Varieties of treatment in the United States, England, West Germany, and France.* New York: Henry Holt and Co., 1988.

Planck, Max. *Scientific Autobiography and Other Papers.* New York: Philosophical Library, 1949.

Poincaré, Henri. *Science and Hypothesis.* New York: Dover Publications, 1952. Edited by Tania Rose. London: Pluto Press, 1999.

Przibram, Karl, ed. *Letters on Wave Mechanics: Schrödinger, Planck, Einstein, Lorentz.* Translation and introduction by Martin J. Klein. New York: Philosophical Library, 1967.

Remarque, Erich Maria. *The Black Obelisk.* Translated by Denver Lindley. New York: Harcourt Brace, 1957.

Richie, Alexandra. *Faust's Metropolis: A history of Berlin.* New York: Carroll & Graf Publishers, 1998.

Rubin, William S. *Dada, Surrealism, and Their Heritage.* New York: Museum of Modern Art, 1968.

Rudin, Harry R. *Armistice 1918.* Hamden, Conn.: Archon Books, 1967.

Schlipp, Paul Arthur, ed. *Albert Einstein: Philosopher scientist.* La Salle, Ill.: Open Court Press, 1949.

Shub, David. *Lenin: A biography.* Baltimore: Penguin Books, 1967.

Skidelski, Robert. *John Maynard Keynes: Fighting for Britain (1937-1946).* New York: Penguin Books, 2002.

Sprat, Thomas. *History of the Royal Society.* Edited by Jackson I. Cope & Harold Whitmore James. St. Louis: Washington University Press, 1966.

Stachel, John, ed. *Einstein's Miraculous Year: Five papers that changed the face of physics.* Princeton: Princeton University Press, 1998.

Stern, Fritz. *Einstein's German World.* Princeton: Princeton University Press, 1999.

Stuewer, Roger H. *The Compton Effect: Turning point in physics.* New York: Science History Publications, 1975.

Tadie, Jean-Yves. *Proust: A life.* Translated by Evan Cameron. New York: Viking, 2000.

Teller, Edward and Judith L. Shoolery. *Memoirs: A twentieth-century journey in science and politics*. Cambridge, Mass.: Perseus Publishing, 2001.

Thomson, J.J. et al., *James Clerk Maxwell: a commemoration volume, 1831-1931*. Cambridge: Cambridge University Press, 1931.

van der Waerden, B.L., ed. *Sources of Quantum Mechanics*. New York: Dover Publications, 1968.

Wheeler, John Archibald and Wojciech Hubert Zurek, eds. *Quantum Theory and Measurement*. Princeton: Princeton University Press, 1983.

Wythe, William. "Einstein Distracted by Public Curiosity." *New York Times* (Feb. 4, 1929), p. 1.

D. OTHER SOURCES

Bohr, Niels. "The Quantum Postulate and the Recent Development of Atomic Theory." *Nature*. Vol 121, No. 3050 (April 14, 1928), pp. 580-590.

— "Discussion with Einstein on Epistemological Problems in Atomic Physics" (1949). In Schlipp, Paul Arthur (1949), pp. 199-241.

Bohr, Niels, Hendrik Kramers and John Slater. "The Quantum Theory of Radiation" (1924). In van der Waerden (1968).

Born, Max. "On the Quantum Mechanics of Collisions" (1926). In Wheeler and Zurek (1983), pp. 52-55.

— "Einstein's Statistical Theories." (1949). In Schlipp, Paul Arthur (1949), pp. 161-177.

Born, Max, Werner Heisenberg, and Pasqual Jordan. "On Quantum Mechanics II" (1925). In van der Waerden (1968), pp. 321-386.

Ehrenfest, Paul. "Adiabatic Invariants and the Theory of Quanta" (1917). In van der Waerden (1968), pp. 79-94.

Einstein, Albert. "On the Motion of Small Particles Suspended in Liquids at Rest Required by the Molecular-Kinetic Theory of Heat." (1905a). In Stachel (1998), pp. 85-98.

— "On the Electrodynamics of Moving Bodies." (1905b). In Stachel (1998), pp. 123-60.

— "Does the Inertia of a Body Depend on Its Energy Content?" (1905c). In Stachel (1998), pp. 161-64.

— "On a Heuristic Point of View Concerning the Production and Transformation of Light." (1905d). In Stachel (1998), pp. 177-98.

— "On the Quantum Theory of Radiation" (1917). In van der Waerden, B.L. (1967), pp. 63-77.

— "Ether and the Theory of Relativity" (1921). In *Sidelights on Relativity*. New York: Dover Publications, 1983.

— "Fundamental Ideas and Problems of the Theory of Relativity" (1923), The Nobel Institute, available as of 10/4/2003 at http://www.nobel.se/physics/laureates/1921/einstein-lecture.html.

— "Field Theories, Old and New." *The New York Times*, Feb. 3, 1929.

— "Living Philosophy" (1931a). In Leach (1931), pp. 3-7.

— "Maxwell's Influence on the Development of the Conception of Physical Reality" (1931b). In Thomson, J.J. et al. (1931), pp. 66-73.

— "How I Created the Theory of Relativity." *Physics Today.* Vol 35, No. 8 (Aug, 1982a), pp. 45-47.

Elon, Amos. "The Wanderer." *New York Review of Books.* Vol. 49, No. 16 (Oct. 24, 2002b), p. 18.

Forman, Paul. "Weimar Culture, Causality, and Quantum Theory, 1918-1927: Adaptation by German physicists and mathematicians to a hostile intellectual environment." *Historical Studies in the Physical Sciences.* Vol. 3 (1971), pp. 1-115.

German Historical Museum. "Maximilian Harden: 1861-1927." Web site: http://www.dhm.de/lemo/html/biografien/HardenMaxilian/.

Hawking, Stephen. "Dirac Memorial Address" (1998). In Goddard (1998).

Heisenberg, Werner. "Quantum-Theoretical Re-Interpretation of Kinematic and Mechanical Relations" (1925). In van der Waerden (1968), pp. 261-276.

— "The Physical Content of Quantum Kinematics and Mechanics" (1927). In Wheeler and Zurek (1983), pp. 62-84.

— "Introduction" in Born (1971).

— "Development of Concepts in the History of Quantum Theory" (1973). In Mehra (1973), pp. 264-275.

— "Commentary." (1967). In Wheeler and Zurek (1983), pp.7-8, 56-57.

Heitler, W. "Erwin Schrödinger." *Biographical Memoirs of Fellows of the Royal Society.* Vol. 7 (1961), pp. 221-225.

Helmholtz, Hermann von. "The Conservation of Energy" (1863). In Bolles (1997), pp. 400-408.

Hollingdale, R.J. "Introduction" in Arthur Schopenhauer, *Essays and Aphorisms.* New York: Penguin Books, 1973.

Kepler, Johannes. "I Admit the Moon Has Seas," 1610. In Bolles (1997), pp. 245-249.

Ketterle, Wolfgang and Marc-Oliver Mewes. "Bose-Einstein Condensates: A novel form of quantum matter." IEEE Newsletter (Dec. 2001).

Klein, Martin J. "The First Phase of the Bohr-Einstein Dialogue." *Historical Studies in the Physical Sciences.* Vol 2 (1970b), pp. 1-39.

Luchins, Abraham S. and Edith H. Luchins. "Max Wertheimer: His life and work during 1912-1919." *Gestalt Theory.* Vol. 7 (1985), pp. 3-28.

Mach, Ernst. "A New Sense" (1897). In Bolles (1997), pp. 22-28.

Marage, Pierre and Grégoire Wallenborn. "The Birth of Modern Physics" (1999b). In Marage and Wallenborn (1999a), pp. 95-202.

— "Physics Prior to the First Council" (1999c). In Marage and Wallenborn (1999), pp. 70-93.

McCormmach, Russell. "Einstein, Lorentz, and the Electron Theory." *Historical Studies in the Physical Sciences.* Vol. 2 (1970), pp. 41-87.

Mehra, Jagdish. *"Satyendra Bose."* *Biographical Memoirs of Fellows of the Royal Society.* Vol. 7 (1975b), pp. 117-138.

Miller, Arthur I. "The Special Relativity Theory: Einstein's Response to the Physics of 1905." (1982). In Holton and Elkana (1997), pp. 80-101.

Miller, Paul D. "Einstein on Verge of Great Discovery." *New York Times* (Nov. 4, 1928), p. 1.

Nordmann, Charles. "With Einstein on the Battle Fields." *Literary Digest*. Vol. 73 (June 3, 1922), pp. 586-592.

O'Connor, J.J. and E.F. Robertson. "Brillouin." Web site: http://www-groups.dcs.st-and.ac.uk/~history/Mathematicians/Brillouin.html.

Overbye, Dennis. "John Archibald Wheeler: Peering through the gates of time." *New York Times*. Feb. 12, 2002, p. F:5.

Pauli, Wolfgang. "Einstein's Contribution to Quantum Theory" (1949). In Schlipp (1949), pp. 147-160.

Planck, Max. "On the Law of Distribution of Energy in the Normal Spectrum, " *Annalen der Physik*. Vol. 4 (1901), pp. 553-563. Translation found on Web site: http://dbserv.ihep.su/~elan/src/planck00b/eng.pdf.

Poincaré, Henri. "Principles of Mathematical Physics." *Scientific Monthly*. Vol. 82, No. 4 (April, 1956), pp. 165-75.

Rosenfeld, Léon. "Commentary" (1971). In Wheeler and Zurek (1983), pp. 57-61.

Rosenthal-Schneider, Ilse. "Presuppositions and Anticipations in Einstein's Physics" (1949). In Schlipp (1949), pp. 129-146.

Royal Netherlands Academy of Arts and Sciences. *The Trippenhuis*. Web site: http://www.knaw.nl/uksite/uk_trip.htm.

Shara, Michael M. [curator]. *Einstein*. New York: American Museum of Natural History, Nov. 15, 2002-August 10, 2003.

Singer, Wendy. "Portfolio: Einstein and Tagore." *The Kenyon Review*. Vol. 23, No. 2 (Spring, 2001), pp. 7-33.

Sommerfeld, Arnold. "To Albert Einstein's Seventieth Birthday" (1949). In Schlipp (1949), pp. 97-105.

Unsigned. "Gothenburg's Jubilee Exposition." *Scientific American*. Vol 128. (June, 1923), p. 391.

Unsigned. "The Gothenberg Exposition." *Science*. (July, 6, 1923; supplement), p. 7.

Unsigned. "Lawrence of Arabia Hides in London." *New York Times* (Feb. 3, 1929), p. 1.

Unsigned. "Marvels at Einstein for His Mathematics." *New York Times* (Feb. 4, 1929), p. 3.

Unsigned. "Gravitational and Electromagnetic Fields." *Science*. Vol. 74, No. 1922 (Oct. 30, 1931), pp. 438-439.

Unsigned. "Physics in Pre-Nazi Germany." *Science*. Vol. 94, No. 2447 (No. 21, 1941), pp. 488-89.

Unsigned, "Le Professeur Henri Bouasse." Melusine Web site: http://melusine.eu.org/syracuse/mluque/bouasse/.

Wilczek, Frank. "A Piece of Magic: The Dirac equation." In Farmelo (2002), pp. 102-130.

Sources

I: A RADICAL FACT RESISTED

1: THE OPPOSITE OF AN INTRIGUER

p. 3, "**photos from that period**": Clark (1971) inserts.

p. 3, "**Victory all along the line**": thanks to Döblin (1961) 36, for this reminder.

p. 3, "**Rathenau had wept**": Elon (2002b) 18.

p. 3, "**now he appealed**": Friedrich (1995) 28-29.

p. 4, "**Max Wertheimer also rode**": Luchins and Luchins (1985) 19.

p. 4, "**treasonable opinions**": Clark (1971) 233-35.

p. 4, "**Once, in 1910**": Bolles (1991) 19-20.

p. 4, "**the emperor's chancellor had said**": Rudin (1967) 15.

p. 4, "**Gropius was on furlough**": Gay (2001) 8-9.

p. 5, "**a plague was taking liar**": Richie (1998) 295.

p. 5, "**a crowd assembled outside the Reichstag**": Luchins and Luchins (1985) 19.

p. 5, "**A third man**": Born (1971) 148-51.

p. 5, "**When Gropius realized**": Gay (2001) 9.

p. 6, "**he renounced his citizenship**": Clark (1971) 48; Frank (1947) 17.

p. 6, "**A wit during prewar times**": Clark (1971) 214.

p. 6, "**opposite of an intriguer**": *Ibid.* 180.

p. 7, "**The Reichstag was a large, heavy**": Richie (1998) 216.

p. 7, "**The Kaiser had hated**": *Ibid.*

p. 8, "**the development . . . of mustard gas**": Kragh (1999) 132.

p. 8, "**Max Born had supervised artillery**": Luchins and Luchins (1989) 12.

p. 8, "**Ernest Rutherford, had duplicated**": Kragh (1999) 134.

p. 8, "**How naive we were**": Born (1971) 148.

p. 8, "**A newspaperman recognized**": Luchins and Luchins (1985) 19.

p. 9, "**Inside the building**": Kessler (1999) 10.

p. 9, "**the students passed a resolution**": Born (1971) 150.

p. 10, "**had taken part in a historical**": *Ibid.* 151.

2: NOT GERMAN AT ALL

p. 12, "**proletarian face**": Dachy (1990) 100.

p. 12, "**pseudo-Italian setting**": Baedeker (1923) 65.

p. 13, "**Baader rose and shouted**": Dachy (1990) 98-99.

p. 13, "**To be against this manifesto**": *Ibid.* 93.

p. 13, "**The Cabaret's role**": *Ibid.* 33.

p. 14, "**What is truly valuable**": Einstein (1931a) 6.

p. 14, "**Logic is complication**": Dachy (1990) 37.

p. 14, "**Invention is not the product**": Pais (1982) 131.

p. 15, "**Say yes to a life**": Dachy (1990) 93.

p. 16, "**It tore Planck's soul**": Heilbron (2000) 83.

p. 16, "**If the enemy has taken from our fatherland**": Kragh (1999) 140.

p. 16, "**the book was beginning a bestsellerdom**": Forman (1971) 32-35.

p. 17, "**What should our attitude be**": Marage and Wallenborn (1999b) 113-114.

p. 18, "**These cool blond people**": Frank (1947) 113.

p. 18, "**he saved his pocket money**": *Ibid.* 21.

p. 19, "**I dreamed I had cut my throat**": Fölsing (1997) 419.

p. 19, "**a paper offering visible proof**": Einstein (1905a).

p. 19, "**Einstein sometimes wondered why**": Pais (1982) 56.

p. 20, "**Robert Brown had brought this seemingly perpetual motion**": Ord-Hume (1977) 200.

p. 20, "**In one short paper**": Born (1949) 165; Pais (1982) 86; Pauli (1949) 150; Whitaker (1996) 89.

p. 20, "**'step,' as Einstein called**": for example, Einstein (1921) 4; Einstein (1953) 5; Einstein and Maric (1992) 32.

p. 21, "**Here, [in Berlin]**": Fölsing (1997) 419.

p. 22, "**a discrete quantity**": Planck (1901).

p. 24, "**without any of the physical causes**": Einstein (1917) 76.

p. 24, "**the young American, Arthur Compton**": Stuewer (1975).

p. 24, "**a general point of view**": Ehrenfest (1917) 79.

p. 24, "***a priori* incredible**": Clark (1971) 273.

p. 24, "**to show his gratitude**": Fölsing (1997) 420.

3: I NEVER FULLY UNDERSTOOD IT

p. 26, **"Berlin's Anhalt station"**: Baedeker (1923) 173.

p. 27, **"The essential in . . . my type"**: Einstein (1949) 33.

p. 27, **"Man tries to make for himself"**: Einstein (1982b) 225.

p. 28, **"In his head he saw"**: *Ibid*. 243-45.

p. 29, **"This is absolute nonsense"**: Clark (1971) 146.

p. 29, **"Ehrenfest confessed wretchedly"**: Pais (1982) 271.

p. 29, **"Mann . . . thought he had a funny idea"**: Kurzke (2002) 303.

p. 30, **"substitute their . . . cosmos"**: Einstein (1982b) 225.

p. 30, **"Brilliant landscape and satisfied citizens"**: Born (1971) 9.

p. 31, **"He occasionally patronized brothels"**: Overbye (2000) 345.

p. 31, **"Marriage is the unsuccessful attempt"**: Highfield and Carter (1994) 210. See also p. 195 (where Einstein resents wife's use of "we") and p. 206 (Einstein's attraction to "common" women).

p. 31, **"Planck recalled . . . Hermann Helmholtz"**: Planck (1949) 15.

p. 33, **"James Joyce, had a court case"**: Ellman (1982) 452.

p. 34, **"I never fully understood it"**: Clark (1971) 276.

p. 34, **"Sommerfeld boasted"**: Pais (1991) 161.

4: INDEPENDENCE AND INNER FREEDOM

p. 35, **"A lost paradise"**: Einstein (1982b) 3.

p. 35, **"Only the most reactionary places . . . demanded . . . passports"**: Fussel (1971); Fölsing (1997) 80.

p. 36, **"*Socialism is here*"**: Kessler (1999) 79.

p. 36, **"Rumor said Einstein had fled"**: *Ibid*. 88.

p. 36, **"Men always need"**: Clark (1971) 232.

p. 36, **"too small to commit . . . stupidities"**: Einstein (1953) 7.

p. 38, **"historians of science sometimes suspect"**: for example, Steuwer (1975) 35.

p. 39, **"Stark, and Stark alone"**: *Ibid*. 31.

p. 39, **"ultraviolet light increases an electrical discharge"**: Clark (1971) 92.

p. 39, **"J.J. Thomson proposed the modern view"**: Pais (1982) 380.

p. 40, **"Thomas Mann . . . resumed . . . composing"**: Kurzke (2002) 298.

p. 41, **"Bohr hired Betty Schultz"**: Pais (1991) 171.

p. 41, **"Leon Brillouin . . . wrote a letter"**: Clark (1971) 279.

p. 41, **"Those countries whose victory"**: *Ibid*. 273.

p. 41, **"anti-Zionist Jews . . . assimilationist"**: Frank (1947) 152.

p. 42, **"Planck was not . . . quick-witted"**: Clark (1971) 145-46; Pais (1991) 228.

p. 42, **"On the very day"**: Kuhn (1978) 97-98.

p. 44, **"One really ought to be ashamed"**: Born (1971) 10.

p. 44, **"science by no means contents itself"**: Meyerson (1985) 252, review by Einstein.

p. 45, "**Conditions . . . will never be enforced**": Born (1971) 10-11.
p. 45, "**Richard Feynman has pointed out**": Feynman (1994) 40.
p. 46, "**Mach, argued**": for example, Mach (1942).
p. 46, "**Poincaré, argued**": Poincaré (1952).
p. 47, "**heralded by . . . Franz Exner**": Forman (1971) 74-75.
p. 47, "**Bohr wrote his old school chum**": Klein (1970b) 20.
p. 48, "**One who often chatted**": Rosenthal-Schneider (1949) 146.
p. 48, "**What he called the 'merely personal'**": Einstein (1949) 5.
p. 48, "**also from my wife**": Born (1971) 11.
p. 48, "**her 'Albertle'**": Frank (1947) 4.
p. 48, "**not necessary for my happiness**": *Ibid.* 180.
p. 49, "**give Jews inner security**": *Ibid.* 491-92.
p. 49, "**Einstein is in Leiden**": Clark (1971) 282.
p. 50, "**Eric Warburg reported**": Chernow (1993) 217.
p. 50, "**I am against nationalism**": Fölsing (1997) 492.

5: A MERCY OF FATE

p. 51, "**Trippenhuis is as heavy**": Royal Netherlands Academy of Arts and Sciences.
p. 52, "**Ehrenfest . . . once considered moving**": Klein (1970a) 293.
p. 52, "**Ehrenfest was excited by . . . Bohr**": Pais (1991) 190.
p. 52, "**Together they had written a book**": Ehrenfest and Ehrenfest (1990).
p. 52, "**Stupid you certainly are not**": Klein (1970a) 305.
p. 53, "**a first rate mind**": Pais (1991) 228.
p. 53, "**Eddington found star displacement**": Clark (1971) 284.
p. 54, "**Walther Rathenau told an acquaintance**": Kessler (1999) 70.
p. 55, "**It is a mercy of fate**": Fölsing (1997) 440.
p. 55, "**answers only. . . . 'Why?'**": Born (1971) 13.

6: PICTURESQUE PHRASES

p. 57, "**Bohr arrived . . . spirit of Father Christmas**": Fölsing (1997) 477.
p. 57, "**Einstein . . . wept when his mother died**": Clark (1971) 243.
p. 57, "**sent a note to his mother**": *Ibid.* 287.
p. 57, "**Neutralia**": Fölsing (1997) 477.
p. 58, "**When Bohr became a professor**": Pais (1991) 166-67.
p. 58, "**forced Einstein to guide Bohr on foot**": *Ibid.* 228.
p. 58, "**Planck's misfortune wrings my heart**": Heilbron (2000) 83-84.
p. 59, "**Bohr coined the name 'correspondence principle'**": Pais (1991) 193.
p. 60, "**magic wand**": Whitaker (1996) 123.
p. 60, "**village would soon build a mosque**": Baedeker (1923) 176.
p. 61, "**Years later Bohr recalled**": Bohr (1949) 206.

p. 63, "**Lorentz, still believed in an ether**": Kragh (1999) 111; Kuhn (1978) 139.

p. 63, "**removal of the late Mr. Hooke's portrait**": Berlinski (2000) 144.

p. 63, "**I shall not here discuss**": Pais (1991) 232.

p. 64, "**in his visit . . . radiation is . . . what he discussed**": *Ibid.*

p. 65, "**Important, if right**": Pais (1982) 154.

p. 66, "**This is an enormous achievement**": Pais (1991) 154.

p. 66, "**Bohr's unique instinct and tact**": Einstein (1949) 45–47.

p. 66, "**I must give up doing physics**": Pais (1991) 189.

p. 66, "**Not often in life**": *Ibid.* 228.

7: SCIENTIFIC DADA

p. 69, "**a German publisher had commissioned Einstein**": Pais (1982) 299.

p. 69, "**more than 100 new ones**": Clark (1971) 307.

p. 70, "**scientific dada**": Fölsing (1997) 462.

p. 70, "**Max von Laue . . . was there**": Clark (1971) 321.

p. 71, "**so far achieved nothing in theoretical physics**": Fölsing (1997) 465.

p. 71, "**you are a little child**": Born (1971) 40.

p. 72, "**most active in Germany**": Kragh (1999) 24.

p. 72, "**a wonderful paper by Lenard**": Einstein and Maric (1992) 36.

p. 73, "**vague heart conditions are as standard a diagnosis**": Payer and White (1988) a theme of the work.

p. 73, "**nickel-silver baskets**": Ford (1947) 37.

p. 73, "**anti-Berlin movement**": Stern (1999) 130.

p. 73, "**Planck was disgusted with Wien**": *Ibid.*

p. 74, "**world is a curious madhouse**": Fölsing (1997) 455.

p. 74, "**with a pencil borrowed at the last minute**": Clark (1971) 327.

p. 74, "**Lenard . . . did not shy from . . . racial cracks**": Frank (1947) 232.

p. 75, "**not yet made it possible to extend the absolute time interval**": Clark (1971) 327.

p. 765 "**he volunteered as an informer**": Jerome (2002) 208.

8: SUCH A DEVIL OF A FELLOW

p. 76, "**Weyland's secret financiers**": Fölsing (1997) 461.

p. 76, "**Einstein and Elsa hosted a dinner party**": Kessler (1999) 155–157.

p. 77, "**Do not believe a word of it**": Clark (1971) 339–40.

p. 77, "**Elsa told Kessler**": Kessler (1999) 156.

p. 77, "**Paracelsus . . . coined pseudoscientific words**": *Oxford English Dictionary (second edition)* (1992) see: alkahest.

p. 77, "**sensing the significance**": Kessler (1999) 156.

p. 78, "**Tagore came to Berlin in 1921**": Pais (1994) 99.

p. 78, "**My husband mystical!**": Clark (1971) 340.

p. 79, "**Kessler . . . was astonished**": Kessler (1999) 138.

p. 79, "**H.G. Wells published a story**": *The World Set Free* (Thanks to Sheldon Rampton for the source).

p. 79, "**You haven't lost anything**": Frank (1947) 173-74.

p. 81, "**Kramers, thought he found a way**": Pais (1982) 238.

p. 81, "**Kramers . . . changed**": *Ibid.*

p. 82, "**tribal companions in Dollaria**": Fölsing (1997) 495.

p. 82, "**being taught by Hebrews**": *Ibid.* 523-24.

p. 82, "**a weekly colloquium**": Frank (1947) 111.

p. 83, "**voted to limit their membership**": Laqueur (2000) 192.

p. 84, "**Ehrenfest sent Einstein a note**": Klein (1970b) 1.

p. 84, "**my most impressive scientific experience**": Born (1971) 75.

p. 84, "**von Laue energetically challenged Einstein**": Klein (1970b) 12.

p. 84, "**you are such a devil**": *Ibid.* 11-12.

p. 84, "**toy dogs of the women**": Clark (1971) 349.

p. 85, "**I suppose it's a good thing**": Fölsing (1997) 512.

p. 85, "**Einstein was rumored to . . . be on the list**": *Ibid.* 522.

9: INTUITION AND INSPIRATION

p. 86, "**riding in an open car**": Feuchtwanger (1995) 116.

p. 86, "**attempt was made against . . . Harden**": German Historical Museum.

p. 87, "**no vibrato**": Friedrich (1995) 172.

p. 87, "**in a creative fit**": Gay (2001) 152.

p. 88, "**party organized as a military force**": Hemingway (1985) 172.

p. 88, "**potshot patriots**": *Ibid.* 174.

p. 88, "**biggest bluff in Europe**": *Ibid.* 255.

p. 88, "**Lenard said the Jew had been 'justly' killed**": Frank (1947) 193.

p. 88, "**Nabokov's father had been slain**": Friedrich (1995) 88.

p. 88, "**the enemy is on the right**": Feuchtwanger (1995) 117.

p. 90, "**Eldorado of erudition**": Born (1971) 67.

p. 90, "**Theoretical [physics] . . . guesswork**": Crowe (2001) 51.

p. 91, "**the moon might be a mirror**": Kepler (1610) 246.

p. 92, "**'divinely bad' speaker**": Pais (1991) 11.

p. 92, "**[Bohr's] carefully formulated sentences**": *Ibid.* 105.

p. 93, "**Heisenberg rose to challenge**": Cassidy (1992) 129.

10: BOLD NOT TO SAY RECKLESS

p. 94, "**A Japanese publisher asked**": Fölsing (1997) 524.

p. 94, "**In September . . . Einstein . . . received word**": *Ibid.* 535.

p. 95, "**services to the theory of physics**": Clark (1971) 363.

p. 95, "**achievement went to . . . Robert A. Millikan**": Stuewer (1971) 73.

p. 96, "**Why not adopt it?**": *Ibid.* 75.

p. 97, "**bold, not to say reckless**": *Ibid.* 73-74.

p. 97, "**he wrote a classic paper**": Einstein (1905d).

p. 97, "**without any visibile means of support**": Stuewer (1971) 75.

p. 97, "**He had done the math**": Einstein (1905d) 180.

p. 97, "**will probably never be replaced**": *Ibid.* 178.

p. 98, "**Though they may begin to lose faith**": Kuhn (1970) 77.

p. 98, "**Copernicus . . . first rejected**": Crowe (2001) v, 82-87.

p. 98, "**it was the greatest honor and joy**": Pais (1991) 229.

p. 99, "**pleased me as much as the Nobel Prize**": *Ibid.* 229.

p. 99, "**The Present Crisis in Theoretical Physics**": Forman (1971) 62.

p. 99, "**At Stockholm he was more direct**": Pais (1991) 470.

p. 100, "**Two days before Einstein's crisis lecture**": Stuewer (1971) 232.

p. 100, "**conversed with Japan's empress in French**": Frank (1947) 198.

p. 100, "**Elsa had something serious to worry about**": Fölsing (1997) 548; Highfeld and Carter (1994) 206; Shara (2002).

p. 100, "**tribal companions**": Fölsing (1997) 530.

p. 101, "**men with a past but without a future**": *Ibid.* 529.

p. 101, "**rooting through garbage cans**": Richie (1998) 323.

p. 101, "**Property was something to hold on to**": see Remarque (1957).

11: A COMPLETELY NEW LESSON

p. 102, "**Bohr received a warning**": Stuewer (1971) 241.

p. 103, "**signed by the Dutch physicist**": *Ibid.* 234.

p. 104, "**the mistaken notion is to get some idea**": *Ibid.* 90.

p. 105, "**my own contribution**": *Ibid.* 160.

p. 106, "**Compton eventually wrote a report**": *Ibid.* 195.

p. 106, "**tree-lined paths**": Personal experience.

p. 107, "**Compton gave his breakthrough lecture**": Stuewer (1971) 235.

p. 107, "**December he reported his discovery**": *Ibid.* 232-3.

p. 107, "**Debye read Compton's report**": *Ibid.* 234.

p. 107, "**Debye had already considered**": *Ibid.* 235-36.

p. 108, "**Sommerfeld . . . saved Compton's fame**": *Ibid.* 247.

p. 108, "**No, Debye insisted**": *Ibid.* 237.

p. 108, "**a dog that did not bark**": *Ibid.* 34-35.

p. 109, "**almost blunted purpose**": *Hamlet* III, iv.

12: SLAVES TO TIME AND SPACE

p. 110, "**Seen from a ferry**": Bennet (1914) 161-63; Fenneberg (1946) 44.

p. 110, "**but kill the patient**": Pais (1991) 228.

p. 111, "**Einstein did not demand**": Einstein and Maric (1992) xviii.

p. 111, "**turn listeners into secretaries**": Pais (1991) 12.

p. 111, "**metaphor of a beer stein**": Frank (1947) 71.

p. 112, "**they had missed their stop**": Pais (1982) 221.

p. 113, "**Einstein published a paper that summer**": Klein (1970b) 16.

p. 113, "**Matchboxes in Denmark**": Fenneberg (1946) 3.

13: WHERE ALL WEAKER IMAGINATIONS WITHER

p. 116, "He made a public appearance at the Prussian Academy": Klein (1970b) 36; Clark (1971) 494.

p. 117, "Rumor had it that Kahr planned to announce": Feuchtwanger (1995) 133.

p. 117, "Planck was especially worried": Clark (1971) 373-74.

p. 117, "Hundreds of paper mills": Richie (1998) 321.

p. 118, "not psychologically incomprehensible to me": Clark (1971) 280.

p. 118, "sent the porter to say": Frank (1947) 163.

p. 119, "You have produced so much young talent": Cassidy (1992) 102.

p. 119, "it is not right for me to take part in . . . Solvay": Clark (1971) 280.

p. 119, "send me no further invitations": Marage and Wallenborn (1999b) 115.

p. 119, "It is unworthy of cultured men": Clark (1971) 281.

p. 119, "Newton's portrait on the wall": Photo of study.

p. 119, "without exception, scientific propositions are wrong": Kessler (1999) 233.

p. 120, "Einstein issued a statement": Clark (1971) 366.

p. 120, "Einstein published a newspaper article": Klein (1970b) 39.

p. 120, "Bohr published a paper, jointly attributed": Bohr, Kramers, and Slater (1924).

p. 122, "he joined Berlin's New Synagogue": Pais (1982) 527; Elon (2002a) 360.

p. 122, "talked and laughed with Abram Joffe": Clark (1971) 380.

p. 122, "Completely negative": Pais (1991) 287.

p. 122, "I would rather be a cobbler": Born (1971) 82.

p. 123, "Heisenberg was introduced to the great man": Cassidy (1992) 179.

p. 123, "I do not see [BKS] as an essential progress": Ibid. 172.

p. 123, "Einstein had a long list of objections": Klein (1970b) 33.

p. 124, "Einstein 'had a hundred arguments'": Cassidy (1992) 179.

p. 124, "You can talk about . . . Buddha, Jesus, Moses": Overbye (2002).

p. 125, "Schrödinger . . . sent Bohr a congratulatory letter": Stuewer (1971) 298.

p. 125, "would not allow themselves to be dispensed": Klein (1970b) 33.

p. 125, "sent . . . Bette Neumann, a note": Pais (1982) 320; Fölsing (1997) 548.

p. 126, "The available data are not sufficient": Pais (1991) 237.

p. 126, "Fritz Haber wrote to Einstein": Ibid.

14: A TRIUMPH OF EINSTEIN OVER BOHR

p. 127, "Let us hear no more about war": Gay (2001) 121; Richie (1998) 333.

p. 127, "Einstein toured a monument": Gay (2001) 97.

p. 127, "The tower's supporters liked": Friedrich (1995) 164.

p. 128, "a stylish newspaper filled with photographs": Forman (1971) 101.

p. 128, "The notion that nature is comprehensible": *Ibid.*

p. 129, "When Rutherford first read Bohr's proposal": Pais (1982) 153.

p. 129, "*the Pirandello play*": Kessler (1999) 247.

p. 129, "Geiger and . . . Bothe . . . went to work": Pais (1982) 237.

p. 129, "Compton was still at work": Klein (1970b) 34.

p. 129, "Einstein was feeling triumphant": Pais (1999) 238.

p. 130, "Physics is very muddled again": Clark (1971) 405-6.

p. 130, "I firmly believe you are right": Klein (1970b) 33.

p. 130, "Einstein . . . was merely unusually lucky": Kessler (1999) 250.

p. 130, "a subtle theme to meditate upon": McCormmach (1970) 64.

p. 131, "Geiger sent Bohr a note": Stuewer (1975) 301.

p. 131, "Bohr replied promptly": *Ibid.*

p. 131, "Just this moment I have received": *Ibid.*

p. 132, "Compton's cloud-chamber results": Klein (1970b) 303-4.

p. 132, "it was a magnificent stroke of luck": Stuewer (1971) 303-4.

p. 132, "We both had no doubts": Klein (1970b) 34.

p. 132, "paradoxical theories . . . paradoxical phenomena": *Ibid.* 35.

p. 133, "require a sweeping revolution": *Ibid.* 33.

II: A RADICAL THEORY CREATED

15: SOMETHING DEEPLY HIDDEN

p. 137, "New plazas, streets, and buildings": Unsigned magazine report (June 1923) 391.

p. 138, "the text he carried in his pocket": Einstein (1923).

p. 138, "the largest vaulted roof ever constructed": Unsigned magazine report (June 1923) 391.

p. 138, "a history of cutting tools": Unsigned magazine report (July 1923) 7.

p. 139, "most profound physical problem": Einstein (1923) 490.

p. 140, "more depth than surface": Cézanne (1941) 490.

p. 141, "this experience made a deep and lasting impression": Einstein (1949) 9.

p. 141, "bona fide scientific knowledge": Einstein (1923) 482.

p. 141, "observable facts can be assigned": *Ibid.*

p. 142, "a catalogue and not a system": Clark (1971) 355.

p. 142, "not sufficiently advanced in our knowledge": Einstein (1923) 484.

p. 143, "The mind striving after the unification": *Ibid.* 489.

p. 143, "the criterion of mathematical simplicity": *Ibid.*

p. 144, "not free from arbitrariness": *Ibid.*

p. 145, "a kind . . . of artistic satisfaction": Helmholtz (1863) 400-1.

p. 145, "accurate observation and searching thought": Mach (1897) 28.

p. 145, "about the famous Mr. Einstein": Kurkze (2002) 523.

16: COMPLETELY SOLVED

p. 146, "I do not have to read the thing": Born (1971) 199.

p. 146, "My friend and colleague M. Besso": Einstein (1905b) 159.

p. 147, "prepared over a period of years": Clark (1971) 115.

p. 147, "laws of motion . . . contradicted demonstrated facts": Einstein (1905b) 123.

p. 147, "In old age, Einstein often said": Pais (1982) 13.

p. 147, "Lorentz-Einstein theory": Kragh (1999) 92.

p. 148, "Einstein's father . . . producing dynamos": Fölsing (1997) 9.

p. 149, "Zeno overlooks part of the story": Gullberg (1997) 276.

p. 150, "Galileo solved this one": Galileo (1967) 186-88.

p. 151, "the same laws of electrodynamics and optics": Einstein (1905b) 124.

p. 152, "Galileo had raised the question": Galileo (1954) 42.

p. 153, "Einstein . . . had imagined what it would be like": Clark (1971) 114.

p. 154, "Lorentz . . . clarified the concept": McCormach (1970) 48.

p. 155, "Serbian nationalists would even argue": Highfield and Carter (1994) 108.

p. 156, "light always propagates . . . with a distinct velocity": Einstein (1905b) 124.

p. 156, "Poincaré told an audience in St. Louis": Poincaré (1956) 174-5.

p. 157, "the absolute speed of light was untenable": Stachel (1998) 111-12.

p. 158, "Lorentz understood what had happened": Pais (1982) 167.

p. 158, "a few physicists had recognized . . . a new Copernicus": Clark (1971) 142.

p. 159, "I've completely solved the problem": Einstein (1982a) 46.

p. 159, "It's a very beautiful piece of work": Highfield and Carter (1994) 114.

p. 159, "He produced the same four equations": Clark (1971) 120.

p. 160, "the electromotive force . . . is a secondary phenomenon": Frank (1947) 64.

17: EXCITING AND EXACTING TIMES

p. 162, "The kind of work I do": Frank (1947) 119.

p. 162, "showed in a convincing manner": Ibid. 217.

p. 163, "Johannes Stark . . . asked Einstein": Einstein (1982a) 47.

p. 163, "Gravity . . . hides crucial information": Pais (1982) 178.

p. 164, "the happiest thought": Ibid.

p. 165, "Barely a quarter of Germany's people lived in cities": Laqueur (2000) 25.

p. 165, "eventually the Nazi party could be voted into power": Ibid.

p. 166, "Einstein . . . knew nothing of the Eötvös experiments": Fölsing (1997) 303.

p. 168,"**the ease with which Einstein would . . . change**": *Ibid.* 283.

p. 168,"**Newton had wondered about**": Pais (1982) 194 and 200.

p. 172,"**Assuming that Newton's value . . . was correct**": Pais (1982) 199–200.

p. 173,"**Einstein feared it might be**": *Ibid.* 200.

p. 176,"**he visited Berlin on an unsuccessful job hunt**": Einstein, *Collected Papers* V, 457 n 4.

p. 176,"**April 30, 1912**": Einstein, *Collected Papers* V.

p. 176,"**In Prague . . . George Pick told Einstein**": Frank (1947) 82.

p. 177,"**perhaps the greatest intellectual stride**": Einstein (1931b) 69.

p. 177,"**May 7, 1912**": Einstein, *Collected Papers* V.

p. 177,"**physical laws without reference to geometry**": Einstein (1982a) 47.

p. 177, "**Poincaré's argument that Euclid's geometry**": Poincaré (1952) a theme of work.

p. 177,"**Johannes Kepler once . . . wrote an equation**": Koestler (1990) 407.

p. 178,"**Grossman, you must help me**": Pais (1982) 212.

p. 178,"**March 23, 1913**": Einstein, *Collected Papers* V.

p. 179,"**Grossman began to study and teach Einstein**": Einstein (1982a) 47.

p. 179,"**Einstein was unsatisfied**": Pais (1982) 228.

p. 180,"**February 1914**": Einstein, *Collected Papers* V.

p. 180,"**I firmly believe that the road taken**": Pais (1982) 245.

p. 180,"**Thomas Mann . . . set his novel aside**": Kurzke (2002) 292.

p. 181,"**a chain of false steps**": Pais (1982) 271.

p. 181,"**they are merely arbitrary abstractions**": Einstein (1949) 67.

p. 181,"**the correct equations appeared**": Einstein (1982a) 47.

p. 181,"**During the last month I experienced**": Sommerfeld (1949) 101.

p. 182,"**I was beside myself**": Pais (1982) 253.

p. 182,"**Something actually snapped**": *Ibid.*

p. 182,"**for four weeks in November 1915**": *Ibid.* 250-57.

p. 182,"**any physical theory**": *Ibid.* 256.

p. 182,"**he told Sommerfeld**": Sommerfeld (1949) 101.

p. 183,"**You will be convinced**": *Ibid.*

18: INTELLECTUAL DRUNKENNESS

p. 184,"**Lehrte Station**": Baedeker (1924) 2.

p. 184,"**bills of enormous denominations**": Hemingway (1985) 282.

p. 184,"**There are no beggars**": *Ibid.* 286.

p. 185,"**Germany will wait in vain**": Eyck (1967) v. 1, 243.

p. 185,"**Germany had the misfortune**": Calaprice (2000) 108.

p. 186,"**some of its members threatened a walkout**": Clark (1971) 355.

p. 186,"**Madame Curie showed her support**": *Ibid.* 354.

p. 186,"**an expert on acoustics and hydrodynamics**": Unsigned website, *Le Prof ... Bouasse.*

p. 186,"**the French spirit . . . will never understand**": Frank (1947) 237.

p. 186, "**Einstein got into a car**": Nordman (1922) 587.

p. 187, "**I do not need wine**": *Ibid.* 589.

p. 187, "**Trade between the once-enemy neighbors**": Liesner (1985) table G-7.

p. 188, "**Maurice de Broglie ... returned from that first Solvay**": Jammer (1989) 245.

p. 188, "**father of Leon**": O'Connor and Robertson.

p. 188, "**synthetic theory of radiation**": Jammer (1989) 247.

p. 188, "**Alhazen ... concluded that light ... consisted of particles**": Clegg (2001) 37.

p. 189, "**a photon as a kind of clock**": Jammer (1989) 247.

p. 189, "**De Broglie ... combined the two equations**": *Ibid.* 248.

p. 191, "**harmony of phase will always persist**": *Ibid.* 247.

p. 191, "**I don't want to be one of those poor people**": Frank (1947) 8.

p. 194, "**Many of these ideas may be criticized**": Jammer (1989) 250.

p. 194, "**price of milk**": Fry (1944) 89.

19: THE OBSERVANT EXECUTRIX

p. 195, "**newspapers in Berlin began leaking**": Clark (1971) 373.

p. 195, "**the new rector had erected a war memorial**": Laqueur (2000) 187.

p. 196, "**climbing out of bed**": Shub (1967) 437.

p. 196, "**Rev ... rev ... rev ... vo ... vo ... vo ... lu**": *Ibid.*

p. 196, "**He had been invited before**": Frank (1947) 147.

p. 194, "**the *Berliner Tageblatt* reprinted the rumors**": Brian (1996) 146.

p. 197, "**interview him for some newspaper**": *Ibid.*

p. 197, "**Vassik, had Down syndrome**": Pais (1982) 409.

p. 198, "**English Puritans also published books there**": www.pilgrimpress.com/about.html.

p. 198, "**I am thinking hopelessly**": Born (1971) 81.

p. 198, "**He's a skeptical fellow**": *Ibid.* 65.

p. 200, "**How great must his faith in the existence of natural law**": Einstein (1982b) 62.

p. 200, "**the ghastly apparition of atheism**": Clark (1971) 502.

p. 201, "**his ambition appears to have been discovering how nature really works**": Hollingdale (1973) 9.

p. 201, "**Herodotus told the story**": Herodotus (1954) 70.

p. 201, "**Chance and caprice rule the world**": La Rouchefoucauld, Maxim, 435.

p. 202, "**the observant executrix of God's orders**": Drake (1978) 225.

p. 203, "**Paul Langevin told of ... de Broglie's ideas**": Jammer (1989) 251.

20: IT MIGHT LOOK CRAZY

p. 204, "**I had no idea that what I had done was really novel**": Pais (1982) 424.

p. 205, "**I do not know sufficient German**": Jammer (1989) 251.

p. 205, "**the first revolutionary contribution . . . in a dozen years**": Whitaker (1996) 127.

p. 205, "**Bose's paper . . . began**": Clark (1971) 409.

p. 205, "**your gas has nothing to do with true light**": Jammer (1989) 251.

p. 207, "**He told the members about Bose's work**": Fölsing (1997) 575.

p. 207, "**Einstein dismissed the Bose-Einstein work**": Pais (1982) 423n.

p. 208, "**The whole modern conception of the world**": Monk (1991) 139-40.

p. 208, "**Mann . . . finally finished his** *Magic Mountain*": Kurzke (2002) 296.

p. 209, "**free from any control by the reason**": Rubin (1968).

p. 209, "**Einstein received a letter from Paul Langevin**": Clark (1971) 408.

p. 209, "**'Read it,' Einstein urged**": Przibram (1967) xiv.

p. 209, "**in his second paper on Bose-Einstein**": Fölsing (1997) 576.

p. 209, "**The first laboratory confirmation**": *Ibid.*

p. 209, "**it took almost 70 more years**": Ketterle and Mewes.

p. 210, "**Einstein read . . . still a third paper**": Fölsing (1997) 857.

p. 210, "**a local reporter proved to be an old acquaintance**": Kantha (1996) 25.

p. 210, "**overestimated the courage and integrity**": *Ibid.* 26.

p. 210, "**a foul smelling flower**": Brian (1996) 152.

21: TAKING NOTHING SOLEMNLY

p. 211, "**Bose had applied to Dacca university**": Brown (1974) 83.

p. 211, "**a postcard written in Einstein's own hand**": Shara (2002).

p. 211, "**The German consulate . . . waived the standard fee**": Brown (1974) 130.

p. 212, "**William Faulkner was there**": Blotner (1974) v. 1, 449.

p. 212, "**Germany's borders had remained unsettled**": Feuchtwanger (1995) 171.

p. 212, "**Brecht . . . was now in Berlin**": Gay (2001) 128.

p. 213, "***Wozzeck* for its world premiere**": *Ibid.* 131.

p. 213, "**Yehudi Menuhin . . . Vladimir Horowitz**": Friedrich (1995) 175.

p. 213, "**I loved the rapid, quick-witted**": Gay (2001) 129.

p. 213, "**How did you discover your method**": Brown (1974) 128.

p. 213, "**By vocation and imagination . . . a great teacher**": Mehra (1975b) theme of essay.

p. 214, "**letters of introduction and commendation**": *Ibid.*

p. 214, "**On my return to India I wrote some papers**": *Ibid.*

p. 214, "Never accept an idea as long as you yourself are not satisfied":
Ibid.

p. 215, "the 50th anniversary of Lorentz's doctorate": Pais (1991) 243.

p. 215, "Queen Wilhelmina attended": Haas–Lorentz (1957) 148.

p. 215, "Lorentz had . . . coined the name 'electron'": McCormach (1990).

p. 215, "Max Born . . . was taking advantage of American lecture fees":
Pais (1991) 285.

p. 215, "'Not necessary,' said . . . James Franck": Jammer (1989) 252.

p. 216, "energy and matter are different forms of the same thing": Einstein
(1905c).

p. 216, "Ehrenfest showed off two of his bright students": Fölsing (1997)
579.

p. 216, "Bohr was dubious of this latest idea": Pais (1991) 243.

22: HOW MUCH MORE GRATIFYING

p. 218, "Arnold Sommerfeld raised the news": Heitler (1961) 222.

p. 219, "Schrödinger was particularly eager": Jammer (1989) 258.

p. 219, "his chief love": Heitler (1961) 224.

p. 219, "he, too, longed for a unified field theory": *Ibid.*

p. 219, "the foam on a wave of radiation": Jammer (1989) 257.

p. 219, "Debye and Schrödinger agreed": *Ibid.* 258.

p. 220, "the thinnest spot in a board": Frank (1947) 117.

p. 220, "He was already well acquainted": Jammer (1989) 257.

p. 220, "electron's angular momentum cannot decrease continuously":
Whitaker (1996) 111. This passage also reflects e-mail discussions with Whitaker.

p. 221, "De Broglie had proposed . . . a meaning": Kragh (1999) 165.

p. 221, "Square dancers will have": thanks to Dr. C. Wiggins for inspiring this
image.

p. 221, "Schrödinger . . . recognized a promising . . . approach": Jammer
(1989) 258.

p. 221, "the result . . . did not match the . . . data": *Ibid.*

p. 222, "It is well known to science historians": for example, Whitaker
(1996) 139.

p. 222, "an idea studied by . . . Debye": Jammer (1989) 263.

p. 222, "establish a much more intimate connection": *Ibid.*

p. 222, "The true mechanical processes": *Ibid.*

p. 223, "those damn jumps": Whitaker (1996) 144.

p. 223, "It is hardly necessary to point out": Jammer (1989) 261.

p. 224, "merely as a convenient means of picturization": *Ibid.* 262.

p. 224, "Classical mechanics . . . breaks down": *Ibid.*

p. 224, "One may, of course . . . be tempted": *Ibid.* 260.

p. 225, "The psi function is to do no more": *Ibid.* 266.

p. 225, "You can imagine the interest and enthusiasm": Przibram (1967) 6.

p. 225, "it would be much more beautiful": *Ibid.* 68.

p. 225, "**Planck pointed your theory out to me**": *Ibid.* 24.

p. 225, "**Your approval and Planck's mean more to me**": *Ibid.* 26.

III: A RADICAL UNDERSTANDING DEFIED

23: SORCERER'S MULTIPLICATION

p. 229, "**A new reality is in the air**": Richie (1998) 333.

p. 229, "**Construction work was peppered**": *Ibid.* 331.

p. 229, "**the rising hemline reached the knee**": Laqueur (2000) 31.

p. 229, "**The style was . . . 'American'**": Richie (1998) 339.

p. 230, "**Hitler won Goebbels to his side**": Friedrich (1995) 199.

p. 230, "**you have made a decisive advance**": Przibram (1967) 28.

p. 230, "**q-number algebra**": Kragh (1999) 163.

p. 231, "**A real sorcerer's multiplication table**": Pais (1991) 317.

p. 232, "**relationships . . . which in principle are observable**": Heisenberg (1925) 261.

p. 232, "**You don't seriously believe**": Cassidy (1992) 239.

p. 232, "**As he recalled it many years later**": *Ibid.*

p. 232, "**the individual empirical fact is of no use**": Einstein (1982) 221.

p. 233, "**the three-man paper**": Born, Heisenberg, and Jordan (1925).

p. 233, "**the joke going around the labs that season**": Edward Condon, private communication, Washington University, 1964.

p. 234, "**the disadvantage of not being . . . visualizable**": Born, Heisenberg, and Jordan (1926) 322.

p. 235, "**revealed to man for the first time**": Forman (1971) 10.

p. 235, "**number mysticism would be supplanted**": *Ibid.* 103.

p. 235, "**a pair of wonderful papers**": Pais (1982) 442.

24: ADDING TWO NONSENSES

p. 237, "**Proust . . . hoped that there was some relation**": Tadie (2000) 415.

p. 237, "**People were pretty well spellbound**": Hawking (1998) 4-5.

p. 238, "**probably of extraordinary scope**": Pais (1992) 282.

p. 238, "**The most important question**": *Ibid.* 270.

p. 238, "**Göttingen physicists . . . into two camps**": Cassidy (1992) 213.

p. 238, "**I have made a great discovery**": Pais (1991) 421.

p. 239, "**optimism from Sommerfeld**": *Ibid.* 163.

p. 239, "**deep truths**": Bohr (1949) 240.

p. 239, "**adding these two nonsenses**": Cassidy (1992) 192.

p. 239, "**Spin had first been proposed in Copenhagen**": *Ibid.* 208.
p. 239, "**all the right answers**": Pais (1991) 302.
p. 239, "**there's one mathematical tool**": *Ibid.*
p. 240, "**opens up a very hopeful prospect**": Cassidy (1992) 209.

25: ADMIRATION AND SUSPICION

p. 241, "**Einstein occasionally played**": Elon (2002a) 360.
p. 241, "**There was room for 3,000 people**": Baedeker (1923) 147.
p. 241, "**he would take refuge in music**": Clark (1971) 140-41.
p. 241, "**Schrödinger's proof took the most peculiar feature**": Jammer (1989) 271-72.
p. 242, "**Schrödinger's wave mechanics was 'disgusting'**": Kragh (1999) 166.
p. 242, "**Pauli, once complained to Born**": Born (1978) 12.
p. 243, "**admiration and suspicion**": Jammer (1989) 372.
p. 245, "**One obtains the answer**": Born (1926) 54.
p. 245, "**I . . . give up determinism**": *Ibid.*
p. 245, "**as Einstein noted at the time**": Einstein (1917) 76.
p. 246, "**and do not take both**": Albert (1992) 11.
p. 247, "**I start from a remark by Einstein**": Pais (1991) 287.
p. 247, "**any closer to making the connection**": Einstein (1917) 77.
p. 247, "**no useful departure**": Einstein (1949) 89.

26: AN UNRELENTING FANATIC

p. 248, "**Bohr had invited Schrödinger**": Pais (1991) 298.
p. 248, "**Bohr was otherwise most considerate**": *Ibid.*
p. 249, "**If all this damned quantum jumping**": *Ibid.* 299.
p. 249, "**not prepared to make a single concession**": *Ibid.* 298.
p. 249, "**It will hardly be possible**": *Ibid.*
p. 250, "**tolerate the slightest obscurity**": *Ibid.*
p. 250, "**Bohr talked . . . almost in a dreamlike . . . manner**": *Ibid.* 299.
p. 250, "**Mrs. Bohr proved as faithful**": *Ibid.*

27: THE SECRET OF THE OLD ONE

p. 251, "**Goebbels arrived in Berlin**": Friedrich (1995) 189.
p. 251, "**it can be told that physicswise**": Pais (1991) 288.
p. 252, "**but an inner voice tells me**": Born (1971) 91.
p. 252, "**You believe in God playing dice**": *Ibid.* 149.
p. 253, "**In the course of scientific progress**": Heisenberg (1971) x.
p. 253, "**In dealing with the task of bringing**": Bohr (1949) 228.
p. 253, "**Born complained that Einstein had not rejected**": Born (1971) 91.

p. 254, "**This rejection was based on a basic difference**": *Ibid.*
p. 254, "**the same argument . . . was used**": *Ibid.* 210.
p. 255, "**If God had made the world a perfect mechanism**": Born (1949) 176.
p. 255, "**Einstein was returning to square one**": Born (1971) 91.

28: INDETERMINACY

p. 257, "**Heisenberg suddenly recalled Einstein**": Cassidy (1992) 239.
p. 259, "**During the oral exam for Heisenberg's doctorate**": *Ibid.* 152.
p. 260, "**the final failure of causality**": Heisenberg (1927) 83.
p. 260, "**Pauli thought it was great**": Cassidy (1992) 233.
p. 260, "**Bohr was displeased with the new paper**": Rosenfeld (1971) 99.
p. 261, "**it is natural . . . to compare**": Heisenberg (1927) 68.
p. 262, "**the 'orbit' . . . comes into being**": *Ibid.* 73.
p. 262, "**Pauli suddenly arrived**": Rosenfeld (1971) 60.
p. 263, "**not so simple as was assumed**": Heisenberg (1927) 83.
p. 263, "**I owe great thanks**": *Ibid.* 84.

29: A VERY PLEASANT TALK

p. 264, "**Lorentz . . . brought the matter to the attention of Albert**": Clark (1971) 281.
p. 265, "**The electric lights were back**": Richie (1998) 330; Gay (2001) 139.
p. 265, "**A new Germany must be forged**": Friedrich (1995) 201.
p. 265, "**He understands as much about psychology**": Kantha (1996) 74.
p. 265, "**Einstein's favorite psychological doctrine**": Einstein (1949) 3.
p. 265, "**Freud's ideas owed much to Schopenhauer**": Hollingdale (1973) 10.
p. 267, "**Could not one maintain determinism**": Pais (1991) 426.
p. 267, "**de Broglie tried to go beyond**": Kragh (1999) 206.
p. 267, "**quantum mechanics is a complete theory**": Jammer (1989) 371.
p. 269, "**The very nature of the quantum theory**": Bohr (1928) 580.
p. 269, "**independent reality in the ordinary physical sense**": *Ibid.*
p. 270, "**a natural generalization of the classical mode**": *Ibid.* 581.
p. 270, "**language refers to our ordinary perception**": *Ibid.* 590.
p. 270, "**We find ourselves here**": *Ibid.*
p. 271, "**Einstein rose to show**": Jammer (1989) 375.
p. 272, "**the central mystery of the quantum revolution**": Albert (1992) book's theme.
p. 272, "**Heisenberg, Pauli, and Dirac . . . insisted**": Jammer (1989) 375.
p. 274, "**begs the question**": Whitaker (1996) 208.
p. 274, "**Heisenberg and Pauli would analyze the experiment**": Heisenberg (1967) 7-8.
p. 274, "**I am satisfied in every respect**": Cassidy (1992) 254.
p. 274, "**BOHR was towering over everybody**": Whitaker (1996) 210.
p. 275, "**Carry on . . . on the right road**": Brian (1996) 164.

30: THE DREAM OF HIS LIFE

p. 276, "She was Toni Mendel": Fölsing (1997) 616.

p. 276, "Einstein collapsed": Clark (1971) 424.

p. 276, "I have behind . . . a wonderful life": Haas-Lorentz (1957) 150.

p. 276, "an enormous affair with astonishing touches": Ibid.

p. 277, "The noblest man of our times": Brian (1996) 164.

p. 277, "Attending funerals is something one does": Frank (1947) 80.

p. 277, "God has put so much . . . that is beautiful": Calaprice (2000) 328.

p. 277, "between a genius and a quack": Skidelski (2002) 40.

p. 278, "read Spinoza": Fölsing (1997) 602.

p. 278, "the show of the season": Kessler (1999) 349.

p. 278, "Lunched with . . . Richard Strauss": Ibid. 346.

p. 278, "I am forced to laze about": Fölsing (1997) 602.

p. 279, "Ehrenfest did not invent quantum mechanics": Teller and Shoolery (2001) 249 n 6.

p. 279, "Gamow turned up at the door of Bohr's institute": Pais (1991) 324.

p. 280, "Schrödinger . . . wrote Bohr a letter": Fine (1996) 19.

p. 280, "Einstein kept up with the Schrödinger-Bohr letters": Przibram (1967) 31.

p. 282, "In June 1928, Max Planck appeared": Fölsing (1997) 603.

p. 282, "the dream of his life": Clark (1971) 494.

p. 282, "a surprise visit to the Einstein home": Miller (1928) conclusion supported by entire article.

p. 284, "flashbulb monkeys": Elon (2002a) 260.

31: THE SADDEST CHAPTER

p. 285, "Dirac submitted a paper": Wilczek (2002) 103.

p. 285, "the pioneering and heroic era": Kragh (1999) 167.

p. 285, "The Dirac equation . . . can . . . compute the magnetism of electrons": Wilczek (2002) 103.

p. 286, "most of physics and all of chemistry": Wilczek (2002) 125.

p. 286, "he would often merely repeat word for word": Hawking (1998) 9.

p. 286, "I was not very much interested": Ibid. 10.

p. 286, "the most logically perfect presentation": Einstein (1931b) 75.

p. 287, "because it predicted negative energy": Clarification provided by Martin Gutzwiller.

p. 287, "the saddest chapter in modern physics": Cassidy (1992) 284.

p. 287, "Pauli . . . turned to writing a utopian novel": Cassidy (1992) 284.

p. 287, "the more important problems": Ibid.

p. 287, "Pauli told Dirac . . . to try his hand": Ibid.

p. 288, "My name is Mr. Smith": Unsigned newspaper report (February 3, 1929) 1.

p. 288, "**I really don't need any publicity**": Frank (1947) 219.

p. 288, "**accept so much that was unimagined**": Kurzke (2002) 311.

p. 289, "**The contrast between the popular estimate**": Einstein (1982b) 4.

p. 289, "**relativity has not fulfilled expectations**": Bohr (1928) 589.

p. 289, "**Were you never taught in school**": Teller and Shoolery (2001) 103.

p. 290, "**Einstein again dispatched Max Planck**": Fölsing (1997) 604.

p. 290, "**Einstein . . . issued a short statement**": Unsigned newspaper report (January 1929) 1.

p. 290, "**The normal printing . . . sold out at once**": Fölsing (1997) 604.

p. 290, "*Herald Tribune* **published a translation**": *Ibid.* 605.

p. 290, "**Eddington wrote Einstein**": *Ibid.*

p. 291, "**article has permanent value**": Wythe (1929) 1.

p. 291, "**characteristics which especially distinguish**": Einstein (1929).

p. 291, "**Einstein slipped off to . . . Wannsee**": Wythe (1929) 1.

32: A REALITY INDEPENDENT OF MAN

p. 293, "**Berlin's communists began their annual parade**": Feuchtwanger (1995) 213.

p. 294, "**the largest communist apparatus in the world**": Richie (1998) 386.

p. 294, "**Elsa, of course, . . . found it**": Clark (1971) 500.

p. 294, "**the house did not . . . have a telephone**": Fölsing (1997) 674.

p. 294, "**To reach the village**": Baedeker (1923) 209.

p. 295, "**Berlin had returned to horrors**": Richie (1998) 390.

p. 295, "**Tagore . . . returned to Germany**": Singer (2001) 9.

p. 295, "**to Einstein's displeasure**": Clark (1971) 503.

p. 296, "**the Apollo of Belvedere**": Singer (2001) 22-23.

p. 296, "**majestic tranquility**": *Ibid.* 11.

p. 296, "**Beauty is in the idea**": *Ibid.* 23.

p. 297, "**whether Truth is independent**": *Ibid.* 24.

p. 297, "**Then I am more religious**": *Ibid.* 25.

33: A CERTAIN UNREASONABLENESS

p. 299, "**the Nazis astonished themselves**": Richie (1998) 396.

p. 299, "**Demonstrations took place across Berlin**": Kessler (1999) 400.

p. 300, "**Einstein proposed one of his . . . contraptions**": Bohr (1949) 224.

p. 301, "**Léon Rosenfeld . . . reported**": Whitaker (1996) 217.

p. 301, "**the Lord is . . . not malicious**": Clark (1971) 513; Pais (1982) 113.

p. 302, "**Einstein . . . helped Bohr work through the calculations**": Whitaker (1996) 219.

p. 303, "**Einstein never saw himself as tragic**": Einstein (1949) shows how he did view himself.

p. 303, "**akin to . . . the religious worshipper**": Einstein (1982b) 227.

p. 303, "**many had opposed their . . . blasphemy**": Sprat (1966) see e.g. Chapter 1.

p. 304, "**I know this business is free of contradictions**": Pais (1982) 459.

AFTERWORD

p. 306, "**builds on ideas Louis de Broglie had proposed**": Gribbon (1998) 177.

p. 306, "**I never thought he was so attached to me**": Fölsing (1997) 688.

ACKNOWLEDGMENTS

Thanks especially to Dr. Christopher Wiggins, assistant professor of Applied Physics and Applied Mathematics, Columbia University, for his patience in reading and commenting on the manuscript while it was in progress and in need of help. Dr. Martin Gutzwiller and Dr. Andrew Whitaker also read the manuscript and offered valuable corrections. Any surviving errors are my own fault, of course, but they made the book better.

My agent, John Thornton, did more than the usual yeoman's labor of a literary agent, providing me with an excellent sounding board when the project was in its early stages. My brother, Harry Bolles, provided another sounding board, while his son, Harry, helped with research.

E-mail allowed me to quickly resolve some puzzles. Thanks especially to Helge Kragh, University of Aarhus, Denmark; Edith Luchins, Rensselaer Polytechnic Institute; Sheldon Rampton, National Association of Science Writers; Michael Ruh, Society for Gestalt Theory; and John Stachel, Boston University.

Index

A

A Moveable Feast (Hemingway), 212
Acceleration, 164-167
Albert, King of Belgium, 264
Alhazen, 188, 192-193
Alpha decay, 279
American Physical Society, 107
Antheil, George, 265
Antimatter, 305
Anti-relativity movement, 69-75, 146,
    154-155, 197, 254
Anti-Semitism, 49-50, 70, 73, 74, 82, 83,
    85, 86, 88, 90, 185, 196, 299
Antisubmarine technology, 8
Archimedes, 20, 38, 129, 150
Aristotle, 74, 109
Armstrong, Louis, 87
Artillery range-finding equipment, 8
Atheists and atheism, 198, 200
Atomic bomb, 79
Atomic spectrum, 60-61, 66, 103-104,
    130, 148, 190
Atoms, 38, 39, 93, 106. *See also* Electrons
Atoms, Bohr's model, 34, 64-66, 90, 91-
    92, 102, 113-114, 115, 129, 130,
    147-148, 216, 220

Authority, 10, 13, 14, 59, 67, 68, 70, 74,
    77, 84, 124, 126, 131, 195, 201,
    202, 218, 233, 296, 299

B

Baader, Johannes, 12-13
Bacon, Francis, 200
Bauhaus school of design, 40
Besso, Michele, 19, 31, 146, 152, 157,
    178, 235
Big Bang theory, 281-282
BKS theory, 120-124, 128, 129, 131-
    132, 139, 203, 210, 235, 238, 245-
    246
Blumenfeld, Kurt, 36, 41, 49
Bohr, Margrethe, 161
Bohr, Niels, 22, 72, 80, 215, 296
    atomic model, 34, 64-66, 90, 91-92,
        102, 113-114, 115, 129, 130, 147-
        148, 216, 220
    Christian X's audience with, 58
    complementarity interpretation,
        269-272, 281, 289, 292, 303, 305-
        306
    correspondence principle, 59-60,
        239, 240

Einstein's dispute with, 57-67, 98-99, 110-115, 145, 146, 151, 152, 158, 161, 253, 261, 268-275, 279, 281, 286, 291, 292, 297, 300-304, 305
escape from Nazis, 89
friends and admirers, 52, 57-58, 91, 118, 129-130
and Heisenberg uncertainty principle, 256-263, 280
as Institute of Theoretical Physics director, 40
as lecturer and teacher, 52, 59-60, 90, 91, 92, 138, 268-269, 279-280, 289
lectures in Germany, 59-60, 90, 91, 99
mathematical limitations, 92-93, 111, 121, 123, 237-238, 242, 255, 269
Nobel Prize and lecture, 95, 98-100, 139, 257
opposition to light-quantum hypothesis, 23, 48, 60-61, 62, 99-100, 102, 103, 104, 106, 107, 110-115, 120-126, 153, 198, 206
peer comprehension and acceptance of theories, 60, 90, 92, 130
personal characteristics, 58, 61, 64, 92, 110, 111, 124-125, 146, 265, 266, 274-275, 289, 299
professional relationships and work habits, 64, 65, 71-72, 93, 239, 248-250, 256, 260-263, 301
quantum model, 60-61, 62, 64, 65, 81, 99, 102, 104, 106, 118, 120-123, 129-130, 131-133, 203, 238
and Schrödinger's equation, 248-250
theoretical approach, 52-53, 59-61, 65-67, 91, 92, 105-106, 112, 120, 129, 153-154, 220-221, 233, 238-240
view of science, 47, 147
Boltzmann, Ludwig, 19-20, 286
Boltzmann distribution, 19-20

Born, Max, 5, 7, 8, 9, 11, 38, 44, 45, 48, 55, 71, 73, 84, 89, 90, 97, 119, 123, 128, 129-130, 203, 209, 215, 230, 233, 234, 237, 242-247, 251-255, 256, 259, 260, 265, 266, 267-268, 271, 280, 286, 306
Bose, Saryendra, 126, 204, 211-214, 215, 231, 246, 248
Bose-Einstein statistics, 204-209, 210, 213, 218, 233, 286
Bothe, Walter, 129, 130, 131
Bothe-Geiger experiment, 129, 130, 131
Bouasse, Henri, 186
Bragg, Lawrence, 104
Bragg, William, 266, 267
Brecht, Berthold, 212, 278
Breton, André, 209
Brillouin, Leon, 41, 42, 187
Brillouin, Louis, 266
Brillouin, Marcel, 187
Brod, Max, 6, 168
Brown, Robert, 20
Brownian motion, 20, 116, 128, 253, 255
Bruno, Giordano, 77, 124
Buek, Otto, 210
Burton, Richard, 14

C

Calculus
  differential, 107, 177, 241-242
  matrix, 233, 238, 239-247, 248, 281
  Riemannian, 179, 289
Cambridge University, Cavendish Laboratory, 103
Causality, and quantum theory, 48, 123, 180, 182, 245, 260, 267
Cézanne, Paul, 140
Chaos, 254
Chaucer, Geoffrey, 200
China, 200

Christian X, King of Denmark, 58, 113
Clemenceau, George, 31
Cloud chamber experiments, 129, 132
Communism, 13, 88, 252, 293, 294
Compass, 140
Compton, Arthur, 24, 102-109, 129,
    131, 132, 165, 197, 198, 199, 265,
    266, 267
Compton effect, 103-109, 110, 112, 116,
    118, 120, 124, 129, 131, 132, 138,
    139, 142, 153, 197, 206, 207, 210,
    233, 235, 238, 245-246, 257, 259-
    260
Conservation of matter and energy, 47-
    48, 81, 112, 120-121, 122, 123,
    125, 128, 129, 132, 183
Coordinate systems, 150, 181
Copernicus, 98, 103, 150, 158, 202, 203
Cornell University, 90
Correspondence principle, 59-60, 239,
    240
Curie, Marie, 79, 118, 119, 186, 212,
    215, 265, 266

D

Dadaism, 12-15, 70, 209
Dante, 200
Darwin, C.G., 47
Darwin, Charles, 33, 107-108, 165, 230
de Broglie, Louis, 126, 188, 189-194,
    195, 197, 198, 203, 205, 209, 211,
    213, 219, 221, 248, 266, 267, 275,
    306
de Broglie, Maurice, 188
de Broglie wave theory, 189-194, 203,
    205-206, 207, 213, 267
de Donder, Theophile, 266
Death in Venice (Mann), 29
Debye, Peter, 41, 89, 103, 107, 112, 218-
    219, 222, 266
Decline of the West, The (Spengler), 16-17
Denmark
    Einstein's travels to, 110-115, 145,
    146-147, 151, 158, 161, 179

Nazi occupation, 58
    women's suffrage, 58
Dirac, Paul, 230, 231, 237, 266, 272, 285,
    287, 289, 306
Dirac's equation, 285-286, 305
Distant parallelism, 282, 291
Dukas, Helen, 277, 306-307
Duino Elegies (Rilke), 87

E

Ebert, Friedrich, 10, 68, 78
Eckart, Carl, 241
Eddington, Arthur, 53, 54, 55, 215, 290
Edward VII, King of England, 217
Ehrenfest, Paul, 24, 29, 41, 51, 52, 54, 66,
    84, 113, 125, 130, 132, 146, 182,
    197, 198-199, 215, 239, 243, 266,
    273, 274, 275, 277, 278, 279, 304,
    306
Ehrenfest, Tatiana, 52, 197
Ehrenfest, Vassik, 197, 306
Einstein, Albert. See also Relativity
    anti-nationalism, 13-14, 18, 35, 36-
    37, 50, 80, 116, 118, 120, 121-
    122, 165, 185-186, 307
    assassination threat, 85, 87, 89, 94,
    117
    and atomic bomb, 79
    in Berlin, post-World War I, 3-11,
    12-13, 15-19, 21-25, 30, 35, 36,
    37, 45, 49
    Bohr's dispute with, 57-67, 98-99,
    110-115, 145, 146-147, 151, 152,
    158, 161, 253, 261, 268-275, 279,
    281, 286, 289, 291, 292, 297, 300-
    304, 305
    in Caputh, 294-295
    celebrity status and influence, 55-56,
    68-69, 70-71, 74, 76-79, 81-82,
    84-85, 94, 100, 197, 204, 205,
    210, 211, 218, 244, 287-288, 290
    debate on quantum completeness,
    268-269, 271-275, 280
    distant parallelism, 282, 291

enemies and detractors, 56, 69-75, 85, 88, 108-109, 154-155, 196-197

equivalence principle, 166-167, 169, 172, 189-190, 207

family and early life, 57, 141, 148, 191, 307

FBI investigation, 75, 307

friends and supporters, 7, 27-28, 31, 52, 54, 57-58, 63, 70, 73, 78, 84, 98-99, 122, 146, 163, 197, 198-199, 247, 277

health problems and convalescence, 277-278, 281, 290

humor, 78, 111-112

as Kaiser Wilhelm Institute of Physics director, 40

as lecturer and teacher, 28-29, 31-32, 38, 50, 82-83, 138, 168, 214

light-quantum hypothesis, 18, 20-21, 22-24, 33, 34, 38, 39, 44, 61, 63, 74-75, 80, 81-82, 85, 95, 96-97, 99-100, 106-107, 110-115, 131, 138, 139, 207, 210, 286

marriage and children, 23, 28, 31-32, 33-34, 48-49, 100, 101, 168, 197, 241, 298, 307

mass-energy equation, 33, 248, 286, 300-301

and matrix calculus, 243-244

music and social life, 87, 89, 93, 230, 241

natural-law view of science, 44-47, 54, 123, 128, 140-141, 152, 200, 201-202, 203, 208, 237, 252, 254, 270, 280, 291, 298, 301, 303-304

Netherlands lectures and honors, 49, 51-55, 212, 214-215

Nobel Prize and lecture, 33-34, 94, 95, 98-99, 110, 137-145

opposition to BKS theorem, 120-124

as pacifist, 4, 6, 7-9, 30, 41, 79, 180, 187, 210, 307

as patent examiner, 31, 54, 70, 99, 164, 167, 168, 186

peer comprehension and acceptance of theories, 6-7, 28-29, 42, 54-55, 57-58, 63, 84, 96, 97, 99, 107, 146

personal characteristics, 3, 6, 7, 16, 26-27, 32-33, 44, 52, 63, 64, 74, 98, 110-111, 187, 230

as philanderer, 23, 26-27, 30, 31, 34, 48, 100-101, 125, 176, 177, 178, 180, 206, 221-222, 254, 276

at Princeton Institute for Advanced Studies, 89, 197, 305, 307

professional relationships and work habits, 38-40, 64-65, 162-163, 166-167, 168, 198-199, 203, 207, 209, 211, 219-220, 230, 231

and Rathenau's murder, 87, 89, 90, 94

scientific identity and internationalism, 35, 37, 40, 54, 80

and Solvay conferences, 79-80, 90, 118, 119, 120, 172, 203, 204, 264-275, 276, 281, 286, 299, 305

and statistical mechanics, 19-20, 111-112, 116, 128, 204-210, 218, 253, 286

and student revolution, 5-11

Swiss citizenship, 6, 12, 18, 23

theoretical approach, 14-15, 20, 24-25, 27-28, 30, 59-60, 63, 65, 67, 92, 97-98, 105-106, 113, 119-120, 141-143, 153, 154, 155-156, 166-167, 231, 238-239, 303

travels and lecture tours, 27-28, 30, 49, 51-55, 76-77, 78-79, 82, 85, 94-95, 99, 100, 101, 102, 106, 122-123, 177, 186-188, 196, 197, 210, 212, 276-277

on Truth, 295-298

unified field theory, 15, 38, 45, 63, 114, 117, 139, 143, 207, 219, 281-283, 284, 285, 287-288, 289, 290-291, 292, 303

World War I, 180-181, 187

and Zionism, 36, 41-42, 49, 50, 82, 101, 116, 121-122, 125

Zürich professorship and lectures,
    12-13, 18, 24-25, 30, 31, 34, 177-
    178, 218
Einstein, Elsa (second wife), 23, 26-27,
    30, 31, 34, 48-49, 57, 62, 69, 76,
    77, 85, 95, 100-101, 111, 125,
    146, 176, 177, 178, 180, 276, 277-
    278, 283, 289, 291, 294, 295, 306
Einstein, Hans Albert (son), 241, 307
Einstein, Mileva (first wife), 23, 28, 31-
    32, 33-34, 72, 94, 154, 159, 277
Einstein Tower, 127
Electric fields, 148
Electromagnetic fields, 152, 285-286
Electromagnetic induction, 148
Electromagnetic theory, 91, 103-104,
    106, 117
    quantum theory and, 39, 80, 83-84,
        96-97, 102-103, 107
    relativity and, 38, 63, 74, 114, 139,
        143, 147, 148, 152, 153, 158, 160-
        161, 164, 174, 181, 183, 287, 290
    speed of light and, 153
Electromotive force, 148, 152, 160, 164
Electrons, 154. See also Atomic theory;
    Photoelectric effect
    Bohr's orbital "jump" theory, 65-66,
        91, 102, 113-114, 147, 215, 220,
        223-224, 225, 238, 260
    collisions, 244
    de Broglie wave theory, 189-194,
        203, 205-206, 207, 213, 267
    Dirac's equation, 285-286, 305
    discovery, 39, 147
    Schrödinger's equation, 219-225,
        233-234, 240, 241-242, 244, 248-
        250, 251, 255, 271, 272, 305
    spin property, 216, 239, 240, 285-286
Eliot, T.S., 87
Elsevir family, 198
Eötvös, Roland, 166
Equivalence principle, 166-167, 169,
    172, 189-190, 207
Eratosthenes, 202
Ether, 154
Euclid, 45, 69, 201

Euclidean geometry, 46, 142, 177, 178,
    234
Exner, Franz, 47
Explanation, 20, 33, 39, 44, 48, 55, 63,
    66, 79, 106, 115, 124, 128-129,
    142, 144, 147-148, 154, 160, 161,
    185, 188, 192, 193, 199, 201-203,
    208, 225, 231, 233, 243- 245, 253,
    257, 286, 296, 299, 303
    versus intuition, 217, 220-221

F

Faraday, Michael, 148
Fascism, 88, 89, 293
Faulkner, William, 212, 279
Faustian Culture, 17
Feynman, Richard, 45
Fizeau, Armand, 153
Fluorescence, 105
Fourier, Baron, 80
Fowler, Ralph, 266
France
    colonialism, 199-200
    as cultural center, 211-212
    Einstein's visit to, 186-188
    postwar relations with Germany, 185,
        186
Franck, James, 215
French Academy of Sciences, 186
French Revolution, 87
Freud, Sigmund, 46, 265

G

Galileo, 8, 38, 91, 104, 109, 150, 151,
    152, 163-164, 198, 200, 202, 255,
    258, 294
Gamow, George, 279-280
Gases, 205-206, 207, 209, 210, 225
Gay, Peter, 87
Geiger, Hans, 129, 130, 131
Germany. See also World War I
    anti-Semitism in, 49-50, 70, 73, 74, 82,
        83, 85, 86, 88, 90, 185, 196, 299

Bad Nauheim, 72-73
Bavarian coup, 117
as cultural center, 212-213, 265, 278, 294
economic problems, 95, 101, 117, 162, 179, 184-185, 194, 208, 212, 265, 294-295
and European isolationism, 17-18, 54, 185-186, 194, 195, 203, 212
Kaiser Wilhelm Institute of Physics, 40
nationalist movement, 88, 118, 119, 129, 137, 195-196, 208, 278, 306
Nazis, 58, 83, 89, 165, 229-230, 236, 251, 294, 299, 305
Notorious Manifesto of the 93, 37, 42
Physical Society, 27
Reichstag, 5, 7, 8-9
revolution, 3-11, 12-13, 15-19, 21-25, 29, 30-31, 35, 36, 37
science in, 16, 17, 69, 71, 72, 73, 79-80, 90, 118, 119, 127-128, 264-265, 294, 306
thug politics in, 86, 88-89, 117, 293-294
Weimar Republic, 229-230, 278
World War I, 3-5, 7-8, 185
Gestalt psychology, 4, 162
Goebbels, Joseph, 230, 251, 265, 294
Göring, Hermann, 30
Gravitational lens, 53
Gravitational mass, 166, 189
Gravity
and acceleration, 164-167
Newton's law, 168
and photon behavior, 127-128, 168-169
relativity and, 163-175, 181, 182, 183, 287
time and, 174-175
Great Britain, colonialism, 199-200
Gropius, Walter, 4, 5, 40, 41
Grossman, Marcel, 74, 178, 179, 181, 183
Gustav V, King of Sweden, 137, 141, 142
Guye, Paul, 266

H

Haber, Fritz, 126, 299-300
Harden, Maximilian, 86
Heisenberg, Werner, 7, 92, 93, 107, 122-123, 124, 125, 203, 206, 221, 230, 231-232, 233, 234, 237-238, 239-240, 242, 243, 244, 248, 253, 256-263, 265, 266, 267, 272, 274, 279, 287, 306
Heisenberg uncertainty principle, 256-263, 267, 269, 272-274, 281, 300, 301, 302, 305
Helmholtz, Hermann, 31-32, 145
Hemingway, Ernest, 88, 184-185, 212
Henriot, Emile, 266
Herodotus, 201
Hertz, Heinrich, 39
Herzen, Edouard, 266
Hidden-variables interpretation, 306
Hitler, Adolf, 30, 70, 82, 117, 208, 229-230, 251, 299
Hooke, Robert, 63
Hoover, J. Edgar, 307
Horowitz, Vladimir, 213
Hume, David, 199

I

Identity and Reality (Meyerson), 291
Imagination, 14, 27, 29, 86, 137, 146, 158, 199, 246, 270, 294
artistic, 13
Bohr's, 126, 255, 256
Einstein's, 27, 126, 167, 255, 303
Heisenberg's, 256
mathematical, 19, 221
poetic, 98
practical, 140
scientific, 15, 27, 67, 98
teaching, 273
India, 200, 214
Inertial mass, 166, 189
Inertial systems, 150
Italy, 198
Fascism, 88
postwar economy, 184-185

J

Japan, Einstein's lecture tour, 94-95, 99, 100, 106, 177
Joffe, Abram, 118, 122
Jordan, Pasqual, 233, 241
Joyce, James, 13, 33, 87

K

Kahr, Gastov von, 117
Kaiser Wilhelm, 4, 7
Kaiser Wilhelm Institute of Physics, 40
Kamerlingh-Onnes, Helke, 51-52
Kepler, Johannes, 6, 109, 177, 200, 231
Kerensky, Aleksandr, 212
Kessler, Harry, 76, 77, 119, 130, 141, 278, 299, 303
Keynes, John Maynard, 277
Knudsen, Martin, 266
Kramers, Hendrik, 81, 106, 107, 110, 118, 120-121, 132, 248, 266
Kronig, Ralph, 239
Kuhn, Thomas, 98

L

La Rochefoucauls, François, 201
Langevin, Paul, 186, 203, 209, 212, 266, 277, 299
Langmuir, Irving, 266
Laser technology, 97
Lavoisier, Antoine, 18, 87, 236, 294
Lawrence of Arabia, 288
Laws of experience, 45-47, 149, 200, 201
Laws of nature. *See* Natural law
League of Nations, 36-37, 212
Leibnitz, Gottfried, 107
Lenard, Philipp, 65, 71, 72, 88, 89, 91, 118, 306
Lenin, V.I., 12, 195-196, 212, 251
Levi-Civita, Tullio, 179
Light, 116-117. *See also* Photoelectric effect; Photons; Quantum theory

as a clock, 174, 190
motion experiments, 192-193
speed of, 152-153, 156, 157, 172-174
testing gravitation effects on, 127-128
visible frequencies, 174
wave-particle duality, 61-62, 132, 143, 167, 188-194, 206, 215
Lindbergh, Charles, 288
Linnean classification system, 32, 33
Lorentz, Geertruide, 276
Lorentz, Hendrik, 17, 24, 27-28, 29, 36, 38, 51, 52, 53, 55, 63, 79, 96, 97, 119, 146, 147, 156, 157, 158, 159, 181, 203, 215-217, 225, 231-232, 239, 246, 264, 265, 266, 267, 271, 276-277, 299
Ludendorf, General, 117
Luther, Martin, 71
Luxemburg, Rosa, 5-6, 145
Lyell, Charles, 236

M

Mach, Ernst, 46, 147, 153-154, 291
Magic Mountain, The (Mann), 29, 40, 180, 208-209
Magnetism, 140
Mahler, Gustav, 70
Mann, Heinrich, 212
Mann, Thomas, 9, 29, 40, 89, 145, 180, 208-209, 212
Many-worlds interpretation, 306
Marx, Karl, 236
Mass-energy equation, 33, 248, 286, 300-301
Matrix calculus, 233, 238, 239-247, 248, 281
Maxwell, James Clerk, 38, 80, 97, 103, 145, 148, 152, 189
Meinhardt, Willy, 276
Mendel, Toni, 276
Mendelev, Dmitry, 91
Mendelsohn, Erich, 127-128
Menuhim, Yehudi, 213

Meyerson, Emile, 291
Michelson, Albert, 153, 166
Miller, Paul, 282-283
Millikan, Robert A., 95-97, 103
Morley, Edward, 153, 166
Mosely, Henry, 90
Motion
    Newton's laws, 170, 219
    and space-time, 149-151, 175
    theories of, 148-149
Mussolini, Benito, 88, 293

N

Nabokov, Vladimir, 88
National Academy of Sciences, 241
National Research Council, 106
Natural law, 44-47, 54, 123, 128, 140-
    142, 144, 145, 152, 153, 159, 198,
    200, 201-202, 208, 237, 252, 254,
    258, 265, 270, 280, 291, 298, 301,
    303-304
Nazis, 58, 83, 89, 165, 229-230, 236,
    251, 294, 299, 305
Negative energy, 287, 305
Netherlands
    Einstein's travels to, 49, 51-55, 197-
        198, 212, 214-215
    free thought and free press, 198
    Royal Dutch Academy of Arts and
        Sciences, 51
    scientific community, 49, 52, 53, 215
    Trippenhuis, 51
    World War I, 36, 53
Neumann, Bette, 100, 125, 206
Newton, Isaac, 38, 45-46, 55, 63, 69, 74,
    80, 107, 109, 158, 159, 188, 209,
    255, 258, 280
Newton's law of gravity, 168, 172, 182
Newton's law of motion, 170, 219
Newtonian mechanics, 45-46, 59-60,
    222-223, 234
Nobel laureates, 38, 51-52, 56, 65, 73,
    95, 98-99, 306

Nordmann, Charles, 186
Notorious Manifesto of the 93, 37, 42

O

O'Connell, Cardinal, 200

P

Painlevé, Paul, 186
Paracelsus, 77
Pauli, Wolfgang, 122, 126, 130, 132, 203,
    238, 239, 241, 242, 260, 262-263,
    266, 272, 274, 287
Paris Peace Conference, 36-37, 41
Periodic table of elements, 91-92, 115
Philosophical Magazine (journal), 204
Photoelectric effect, 72, 89, 189
    and causality, 48
    discovery, 39
    Einstein's law, 18, 20-21, 22-24, 33,
        34, 38, 39, 44, 61, 63, 74-75, 80,
        81-82, 85, 95, 96-97, 99-100,
        106-107, 110-115, 131, 138, 139,
        207, 210, 286
    Lenard's theory, 65, 71, 74-75, 96
    Maxwell's wave theory and, 97
    Planck's formula and explanation, 18,
        21-22, 42-44, 246
    temperature and, 96, 99
    in X-ray scattering, 105
Photons
    de Broglie wave theory, 189-194,
        203, 205-206, 207, 213, 267
    and technological revolution, 139-
        140
Physical Review (journal), 107
Picasso, 212
Piccard, Auguste, 266
Pick, George, 176
Planck, Emma, 58
Planck, Grete, 16, 53, 58
Planck, Karl, 16, 53

Planck, Max, 16, 18, 19, 27-28, 30, 31-32, 37-43, 46, 47, 49, 53, 55, 56, 58-59, 72, 73, 74, 79, 90, 96, 97, 99, 107, 117, 119, 125, 146, 203, 205, 225, 234, 249, 264, 265, 266, 290, 303
Planck's constant, 22, 42, 44, 255, 258
Planck's law, 39, 42-43, 73, 79, 99, 123, 205, 206-207, 213, 246, 258, 300
Plesch, Janos, 122, 277
Poincaré, Henri, 46, 55, 79, 90, 156, 158, 177, 188
Priestly, Joseph, 86-87
Princeton Institute for Advanced Studies, 89, 197, 305, 307
Principia (Newton), 38
Proust, Marcel, 28, 87, 177, 236-237
Prussian Academy of Sciences, 16, 83, 116, 118, 120, 182, 207, 209, 210, 231, 276, 290
Ptolemy, Claudius, 90-91, 109, 168, 201, 203
Puritans, 198

Q

q-Number algebra, 230, 231, 241-247, 248, 257, 280
Quantum collapse, 272-273
Quantum statistics, 128, 241-247, 251-255, 256, 259, 260, 271, 279, 281, 286
Quantum theory. *See also* Photoelectric effect; Photons
    bedspring analogy, 43-44, 206
    Bohr's atomic model, 60-61, 62, 64, 65-66, 81, 93, 99, 102, 104, 106, 118, 120-123, 129-130, 131-133, 203, 238
    Bohr-Kramers-Slater (BKS) theorem, 120-124, 128, 129, 131-132, 139, 203, 210, 235, 238, 245-246
    Born's probabilities, 242-247, 251-255, 256, 259, 260, 271, 281, 286

and Bose-Einstein statistics, 204-209, 210, 213, 218, 233, 286
Bothe-Geiger experiment, 129, 130, 131
cloud chamber test, 129, 132
complementarity interpretation, 269-272, 281, 289, 292, 303, 305-306
completeness debate, 268-269, 271-275, 280
Compton effect, 103-109, 110, 112, 116, 118, 120, 124, 129, 131, 132, 138, 139, 142, 153, 197, 206, 207, 210, 233, 235, 238, 245-246, 257, 259-260
and conservation of energy, 47-48, 81, 112, 120-121, 122, 123, 125, 128, 129, 132
de Broglie wave theory, 189-194, 203, 205-206, 207, 213, 267, 281
Einstein-Bohr dispute, 57-67, 98-99, 110-115, 145, 146-147, 151, 152, 158, 161, 253, 261, 268-275, 279, 281, 286, 291, 292, 297, 300-304, 305
and electromagnetic wave theory, 39, 80, 83-84, 96-97, 102-103, 107
and gas thermodynamics, 210
gas-radiation analogy, 205-206, 207, 209, 210, 225
ghost-wave notion, 61-62, 120-121, 123, 190, 191, 244-245, 247, 251, 267
and gravity, 168-169
and Heisenberg uncertainty principle, 256-263, 267, 269, 272-274, 281, 300, 301, 302, 305
many-worlds interpretation, 306
Newton-Hooke dispute, 63
ocean-wave analogy, 75, 104-105
opposition to light-quantum hypothesis, 23, 38-39, 48, 60-61, 62, 95, 99-100, 102, 103, 104, 106, 107, 110-115, 120-126, 129, 131, 132-133, 153, 189, 198, 206

Planck's law, 39, 42-43, 73, 79, 99, 123, 205, 206-207, 213, 246, 258, 286, 300
  popular discussion of, 128
  relativity theory linked to, 34, 189-194, 216, 271, 281, 285-286, 302
  Schrödinger's equation, 219-225, 233-234, 240, 241-242, 244, 248-250, 251, 255, 271, 272, 286, 305
  and space-time continuum, 114-115, 116, 117, 131, 139, 235
  tests of, 80-81, 83-84
  train example, 169
  virtual field, 120-121, 125, 271
  wave packets, 61-62, 132, 220, 222, 224, 244, 249
  Wien's work, 73

R

Rathenau, Walther, 3-4, 54, 76-77, 85, 86, 87
Reality, 17, 19, 20, 32, 34, 44-47, 159, 229, 234, 244-245, 253-255, 269, 291, 295, 296-297, 302, 303
Red shift, 174, 190
Relativity
  anti-relativity movement, 69-75, 146, 154-155, 197, 254
  baseball example, 156-157, 159-160
  Besso's contribution, 31, 146, 152, 156, 157, 159
  and Bohr's electron theory, 147
  comprehension and acceptance of, 6-7, 28-29, 42, 54-55, 63, 74, 77-78, 95, 116, 158, 186, 210, 219
  and conservation of energy, 183
  Einstein's defense of, 71-75
  and electromagnetic theory, 38, 63, 74, 114, 139, 143, 147, 148, 152, 153, 158, 160-161, 164, 174, 181, 183, 287
  general theory, 29, 38, 40, 53-55, 63, 71, 97, 114, 143, 162-167, 171, 172-183, 187, 219, 245, 281, 282, 287, 302

  and gravity, 163-175, 181, 182, 183, 287
  hidden-variables interpretation, 306
  Lenard-Einstein debate, 72-74
  Lorentz's work and, 147, 154-156, 157, 158, 159, 181, 271
  mathematics of, 159, 178-179, 180, 181-183
  and Newton's laws, 59-60, 152, 168, 172, 182
  principles, 151-152, 156, 175
  proof of, 48, 53-55, 57, 74, 83-84, 171, 182, 283-284
  quantum theory linked to, 34, 189-194, 216, 271, 281, 285-286, 302
  and Riemann no-parallelism, 282
  simplicity of, 54-55, 159
  and space-time continuum, 114-115, 139, 149-151, 156-158, 174-175
  special theory, 143, 166, 167, 172, 182, 186, 223, 285-286, 301
  and speed of light, 156, 157, 173-174
  sum-over-histories interpretation, 306
  train examples, 27, 153, 163, 165 171, 172
Rembrandt, 198
Ricci, Gregorio, 179
Richardson, Owen, 266
Riemann, Bernhard, 178
Riemannian space, 178, 282, 289
Rilke, Ranier Maria, 87
Rosenblueth, Felix, 36, 41
Rosenfeld, Léon, 301
Rosenthal-Schneider, Ilse, 48, 89
Royal Astronomical Society of London, 55
Royal Society of London, 55, 63
Russell, Bertrand, 9, 94
Russia, 195-196
Rutherford, Ernest, 8, 38, 39-40, 64, 99, 103, 129, 130, 248, 277

S

Saint Augustine, 74
Schönberg, Arnold, 212-213

Schopenhauer, Arthur, 265
Schrödinger, Erwin, 89, 125, 218-225,
    230, 231, 233, 235, 239, 240, 242,
    248-250, 261, 266, 267, 278, 280,
    281, 285, 306
Schrödinger function, 224, 245, 249,
    260, 286
Schrödinger representation, 222, 223
Schrödinger's equation, 219-225, 233-
    234, 240, 241-242, 244, 248-250,
    251, 255, 271, 272, 286, 305
Schultz, Betty, 41
Scientific intrigues and factionalism, 41
Serkin, Rudolph, 87
Shakespeare, William, 98, 177, 208
Slater, John, 118, 120-121, 132
Socialism, 36, 293
Society of French Physicists, 186
Society of German Scientists and
    Physicians, 72
Soldner, Johann, 168
Solovine, Maurice, 186
Solvay conferences, 281, 285
    first (1911), 79-80, 90, 172, 187-188
    third (1921), 79-80
    fourth (1924), 118,119, 203, 204
    fifth (1927), 264-275, 276, 286
    sixth (1930), 299-302, 305
    seventh (1933), 305
Solvay, Ernst, 89-90
Sommerfeld, Arnold, 34, 38, 60, 90, 102-
    103, 104, 108, 119, 147, 181-183,
    203, 218, 239
Sonar, 8
Sonnets to Orpheus (Rilke), 87
South America, Einstein's tour of, 210
Soviet Union, 252
Space-time, 114-115, 116, 117, 131, 139,
    149-151, 156-158, 174-175, 235
Sparticists, 30-31
Spengler, Oswald, 16-17
Stalin, Joseph, 196
Stark, Johannes, 38, 39, 40, 56, 71, 91,
    108-109, 163, 168, 209, 306
Statistical mechanics, 19-20, 111-112,
    116, 128, 204-210, 218, 253, 286

Statistical randomness, 47
Stein, Gertrude, 212
Strauss, Richard, 70, 278
Stravinsky, Igor, 212
Sum-over-histories interpretation, 306
Superconductivity, 52
Surrealist Manifesto, The (Breton), 209
Switzerland
    Caberet Voltaire, 13
    post-World War I intrigue, 30-31

T

Tagore, Rabindranath, 78, 295-298
Teller, Edward, 279
Thales of Miletus, 200-201, 202
Thermodynamics, statistics of, 52
Thomson, J.J., 39, 154
Treaty of Locarno, 212
Treaty of Versailles, 45, 49, 212, 294
Trotsky, Leon, 31

U

Ultraviolet light, 39
Ulysses (Joyce), 87
Understanding, 18-19, 45, 62, 77, 98,
    108, 120, 140, 141-142, 202, 216,
    231, 234, 235, 247, 256, 270, 291,
    296
Unified field theory, 15, 38, 45, 63, 114,
    117, 139, 143, 207, 219, 281-283,
    284, 285, 287-288, 289, 290-291,
    292, 303
University of Berlin, 195
University of Chicago, 95
University of Dacca, 204, 211
University of Göttingen, 28-29, 89, 99,
    119, 122-123, 218, 230, 238, 239,
    251
University of Heidelberg, 118
University of Zürich, 12-13, 18, 24-25,
    30, 31, 34, 177-178, 218, 277

V

van Vleck, John, 132
Verschaffelt, Emile, 266
von Laue, Max, 70, 84, 104, 119, 146

W

Wallace, Alfred, 107-108, 230
Warburg, Eric, 50, 76
Washington University (St. Louis), 102, 106
Wasteland, The (Eliot), 87
Watson, Charles, 236
Wave packets, 61-62, 132, 220, 222, 224, 244, 249
Wave-particle duality, 61-62, 132, 143, 167, 188-194, 206, 215
Wells, H.G., 79
Wertheimer, Max, 4, 5, 8, 11, 162
Weyl, Hermann, 73-74
Weyland, Paul, 70, 75, 76
Wien, Wilhelm, 73, 128, 203, 234, 235, 259
Wilhelmina, Queen of the Netherlands, 215
Wilson, Charles, 266
Wilson, Woodrow, 31, 36-37
Wittgenstein, Ludwig, 208
"Workarounds," 157, 201
World War I, 180-181
    Allied blockade, 10, 41, 49, 212
    Battle of Verdun, 7
    Manifesto to Europeans (protest), 210
    Notorious Manifesto of the 93, 37, 42
    Paris Peace Conference, 36-37, 41
    postwar Berlin, 3-11, 29-30
    western front, 186-187
Wozzeck (opera), 213

X

X-rays
    Compton effect, 102-109, 110, 112, 115, 116, 118, 124, 142
    wave theory, 102

Y

Young, Thomas, 80, 188-189, 192-193

Z

Zangger, Heinrich, 277
Zeitschrift für Physik (journal), 107, 205, 260
Zeno's paradox, 148-150, 159
Zionist movement, 36, 41-42, 49-50, 82, 101, 116, 121-122, 125
Zürich Polytechnic School, 18